COMPUTER METHODS IN GEOLOGY

T. V. LOUDON

Institute of Geological Sciences
London

1979
ACADEMIC PRESS
London · New York · San Francisco

A Subsidiary of Harcourt Brace Jovanovich, Publishers

ACADEMIC PRESS INC. (LONDON) LTD
24–28 Oval Road,
London NW1

United States Edition published by
ACADEMIC PRESS INC.
111 Fifth Avenue,
New York, New York 10003

Library of Congress Catalog Number: 78–18025
ISBN: 0–12–456950–1

Printed in Great Britain by
C. F. Hodgson and Son Ltd
London

COMPUTER METHODS IN GEOLOGY

PREFACE

The geologist has the whole of this planet for his subject matter. From his observations, by intuition and reasoning, he discerns order and pattern, which he expresses as geological description and theory. The mathematical geologist goes one step further, seeking a generalized representation of geological thought, in which relationships are consistent and logical in an abstract sense, regardless of the concrete or substantive objects to which they refer. The systems analyst examines geological activities to determine how computer methods can best contribute. When he has defined the computer procedures, it is the task of the programmer to represent these as computer programs, which control the computer in performing the desired sequence of operations. The user can call on a wide variety of existing programs, and provided he understands the method and can interpret the results, he need not concern himself with the details of the coding. The geologist, mathematical geologist, systems analyst and programmer may all, in fact, be the same individual, but even with specialized support the geologist must understand at least some aspects of all these disciplines to take advantage of computer methods.

Much of the power of computer techniques lies in their use of mathematical methods which place specific problems in a general context. The choice of food in a restaurant menu could have been determined by a computer program which would also help in a study of minerals crystallizing from a molten magma. But, though this may emphasize the unity of science, it has led to the introduction of terms and concepts with which few geologists are familiar. There is thus a barrier. A geologist may be aware of the power of computers and their ability to help him in his work. He may nevertheless be unable to see clearly where their contribution might come, and be unaware of the services that exist to help him.

It is for such an individual that this book was written. The intention is not to explain detailed techniques, but rather to encourage the reader to examine his own working methods, and to develop criteria to guide him in applying the computer. Many text books are available which explain how computers work, how to write a program, or describe methods of data analysis in geology or related subjects. Some of these are listed in the references section, and others can be found by browsing in a library or bookshop. There are also many manuals describing the use of specific programs or computer installations. It is hoped that from this book the reader will gain background knowledge which will assist him in using the computing literature for his own purposes, and to make effective use of his available computing facilities. In short, its aim is to help the geologist to make the computer work for him.

The following individuals are in no way to blame for the contents of this book, except that without them it would never have been written: Professor W. C. Krumbein, Professor P. Allen, Professor D. F. Merriam, Professor E. K. Walton. I am indebted to Dr. W. A. Read and Dr. D. G. Farmer for their helpful comments, to the Director of the Institute of Geological Sciences for permission to publish this work, and to numerous colleagues who have taught me more than I know. I hope it may bring pleasure to them and profit to the rest of us.

CONTENTS

Preface v

CHAPTER 1. ORIENTATION 1

INTRODUCTION

1.1. Comparison of human skills and the abilities of the computer.. .. 1
1.2. Reasons for quantifying, abstracting and generalizing, and their link with computer methods 3
1.3. Quantitative and non-quantitative applications; process control; graphic methods 4
1.4. Systems analysis; systems and models 6
1.5. The scientific method in geology; the organization of this book .. 9

HARDWARE

1.6. Problem solving by geologist and by machine 12
1.7. Data management and numerical analysis 13
1.8. Hardware and software 14
1.9. The computer functions of recording, storing, processing and communicating 14
1.10. Data preparation devices and methods; punch cards, paper tape and magnetic tape; redundancy and error checking 15
1.11. Balance, compatibility and trade-off in data recording; key to disk or cassette tape; direct data entry 21
1.12. Data capture at source 23
1.13. Digitizing graphic data 25
1.14. Representation of data within the computer; tapes, disks and main memory 27
1.15. Bits, BCD and binary arithmetic 30
1.16. Bytes and words; integer and floating point; registers and addresses .. 31
1.17. Output devices and methods; the line printer, graph plotter and other devices 32
1.18. Multiprograming; the buffer and the queue 35
1.19. Remote access; the minicomputer; the desk calculator 36

SOFTWARE

1.20. The interface between hardware and software; machine language .. 37
1.21. User- and problem-oriented languages; assemblers and compilers; compatibility and standards 37
1.22. Programs, subroutines, instructions 38
1.23. Outline of the FORTRAN language.. 39

1.24. Words, locations, addresses; names of variables and arrays 39
1.25. Symbols and expressions; statements; loops and branches 40
1.26. Routines; modular programming 43
1.27. Input and output statements; format statements 45
1.28. Flowcharts 49
1.29. Operating systems and job control languages 50
1.30. Program libraries, packages and systems 51

CHAPTER 2. OBSERVATION AND REPRESENTATION 53

2.1. The objectives of data collection and recording; data as illustration
     and as evidence 53
2.2. The effect of computer methods on data collection 54
2.3. Set theory; sets, subsets, union and intersection 55
2.4. Properties and relationships; formal logic 56
2.5. Data structures; arrays, matrices, trees and networks 58
2.6. Types of data element; quantitative and qualitative 62
2.7. Sampling theory; random, multilevel and sequential sampling .. 64
2.8. Properties of a sample from a normal distribution; sampling design .. 71

REPRESENTATION OF DATA FOR THE COMPUTER

2.9. Constraints imposed by the computer 78
2.10. Potential uses of computer data 81
2.11. Files and subfiles; data sheet design 86
2.12. Fields and records 88
2.13. Input, exchange and storage formats 88
2.14. Coding and abbreviation 92
2.15. Vocabulary control; the thesaurus 98
2.16. Data management; the data library, data base and computable model 99

CHAPTER 3. INTERPRETATION 109

3.1. Objectives in interpreting data 109
3.2. Frame of reference 110
3.3. Redundancy 111
3.4. Transformation and relationship 112
3.5. Distance and metric; geometric and algebraic operations 113
3.6. Weighting of variables 119
3.7. Variables in metric space 120
3.8. Random variables, frequency distributions, descriptive statistics .. 121
3.9. Standardized data, error checking, data transformation 122
3.10. Additional descriptive statistics; efficient and unbiased statistics .. 129
3.11. Cluster analysis 130
3.12. Correlation; principal component analysis 133
3.13. Geometric data processing 138
3.14. Analysis of orientation data 147
3.15. Regression; trend surfaces 149
3.16. Analysis of qualitative data 154
3.17. Sequence and spatial analysis 159
3.18. Symbol plotting and value posting; geographic transformation .. 160
3.19. Moving averages; grey-scale maps; smoothing and filtering 164

3.20. Transition probabilities 167
3.21. Serial correlation; correlograms 170
3.22. Slope analysis 173

CHAPTER 4. EXPLANATION 175

4.1. The nature of explanation; the place of the computer 175
4.2. Probability, simulation, uncertainty and risk 178
4.3. The normal distribution model 184
4.4. Statistical inference; tests of significance and the null hypothesis .. 185
4.5. Causality and correlation 189
4.6. Hypothesis testing.. 189
4.7. Computer simulation; random numbers 191
4.8. Testing subjective judgment; retrospective prediction.. 194
4.9. Interaction of variables; sensitivity analysis 195

THE SYSTEMS APPROACH TO MODELLING

4.10. Integration of ideas in a computer model 197
4.11. Analysis of the system 198
4.12. Expressing implicit information; table lookup and cross-reference;
      generated data 198
4.13. The allocation model; linear programming 200
4.14. Contouring; the spatial model 202
4.15. Sequential data and the transition model 208
4.16. Linking sequences and surfaces 211
4.17. The heuristic approach; the classification model 212
4.18. The process–response model; cybernetics; simulation 214
4.19. A framework for the systematic use of expert opinion; the risk
      analysis model 221

CHAPTER 5. COMMUNICATION 227

5.1. Communication in computer applications 227
5.2. The shared coding scheme, standards and interfaces 227
5.3. Information flow and the system of geological communication .. 229
5.4. Communicating with the computer; interactive and batch processing 231
5.5. Terminals, data links and networks 233
5.6. Documentation; exchange of programs and data 238
5.7. Communication between programs 240
5.8. Communication between the geologist, systems analyst and
      programmer 242
5.9. Word and text processing; the teleconference; the Delphi 242
5.10. Information systems; catalogs and indexes 244

CHAPTER 6. IMPLEMENTATION 247

6.1. Defining the objectives 247
6.2. Review of the existing system and available resources 248
6.3. The system input and output 250
6.4. The driving forces 252
6.5. Cost-effective implementation; hardware, software and methods .. 252

6.6. Organizational aspects   257
6.7. Recapitulation   258

References   261

Subject Index   265

1. ORIENTATION

INTRODUCTION

1.1. Comparison of human skills and the abilities of the computer

This book is about geology, but not about rocks. It is concerned with a tool, the computer, with the techniques of using it, and with the part which it can play in the work of the geologist and the development of his science. It is not always easy to recognize the areas in which computer methods can most effectively be applied. On the one hand, effort may be wasted in programming a computer to carry out a task which could be done in half the time at less cost by conventional methods. On the other hand, important properties of the data may be ignored and unnecessary manual work undertaken through a failure to recognize and apply new techniques.

To illustrate this point and put the computer's contribution into perspective, it may be useful to consider the activities of a well equipped team of geologists on a detailed field survey. Such a team can call on a wide range of skills and experience. They will not only have previously examined many outcrops elsewhere, but may also have studied rock samples and cores from both deep and shallow boreholes. Members of the team may have read detailed descriptions of rocks and fossils and of geological events believed to have occurred in the remote past, and have observed and experimented with geological processes taking place at the present day. A geomorphologist may have seen land forms in the Antarctic and in the Sahara Desert, may have walked and flown over sand dunes and crawled through tunnels within a glacier. He may have seen films of avalanches, land-slips and mud-slides, and cut sections through contorted beds of slumped clay. He is likely to be familiar with the interaction of landscape and its cover of vegetation and to know the broad context in which various types of land form develop. In an area that is new to him, he can see much that he understands and much that is familiar.

A group of geologists can not only call on a formidable background of experience, but can also bring remarkable powers of observation to bear on a problem. A human being can see with his naked eye an extensive landscape,

1

and a tiny grain of sand. With hand lens, optical and electron microscope he can observe minute detail. With binoculars, air and satellite photographs he can extend the range of his observation until he sees half the earth at a glance. With hammer, spade, bulldozer and drill, he has access to fresh material below the weathered layer. With diamond saw and lap he can obtain a precisely smooth surface or a microscopically thin section. The geologist can obtain photographs and interpret patterns not only of visible light but of a wide spectrum of electromagnetic waves. He can, by infrared photography, detect minor variations of temperature in a distant landscape. The geophysicist can estimate the size, shape and nature of buried rock masses from physical properties and can map buried strata by timing shock waves. The geochemist can rapidly and accurately determine the chemical composition of a rock specimen or of a minute spot on a thin section.

When one considers such a range of skills and experience, it at first seems quite unrealistic to suppose that the computer could take over any appreciable part of these activities. Perhaps the most striking characteristic of the geologist when compared to a computer is the enormous amount of information which he has at his disposal. Much of it is obtained through direct observation by eye. The eye is an intricate mechanism, an extension of the brain itself, which can process as well as obtain visual information. The ability to recognize and analyze patterns in space, both on the ground, and in graphic representations such as maps, diagrams and cross-sections, is an important attribute of the geologist which no machine can emulate. The ability to distinguish familiar faces at a glance is better developed in a young child than in the most advanced computer program.

Information held in the human brain is highly structured. Individual items of information are linked by a complex network of associations, so that a single observation may spark off a chain of related ideas and remembered concepts. Observation and hypothesis may be closely linked in a geologist's mind, forming a coherent and integrated framework into which new facts and ideas can be fitted, and where anomalies and contradictions are revealed. Again, no machine can compete with this power to synthesize and relate diverse and often diffuse information. The geologist deals as a matter of course with characteristics of his own observations which he could not describe to a fellow geologist, let alone to a computer.

But, whereas the human being scores in the quantity, complexity and diffuseness of the information which he can handle, the computer scores in the precision with which it can store data, the speed with which it can manipulate them, and the reproducibility of the results. The computer can perform intricate mathematical calculations, can rearrange or select from a file of data at a fraction of the time and cost of the corresponding manual operation.

For examples, the fictitious survey of an area introduced above will again serve.

1.2. Reasons for quantifying, abstracting and generalizing, and their link with computer methods

There are inevitable difficulties in integrating the observations of individuals, with different backgrounds and training, into a single coherent account of an area. A small team is faced in an acute form with some of the problems of any descriptive science. Ideas that cannot be substantiated, and impressions and observations which cannot be described, are of limited value. A stratigrapher may require paleontological identifications and the results of a geophysical survey in order to define the distribution of a particular formation. A geomorphologist might need to know the pattern of tectonic deformation of an area in order to explain the land forms, and geochemical results might find an explanation in a petrological investigation. For these purposes, precise and unambiguous communication is essential. Each member of the hypothetical team must attempt to work to common standards, use the same terminology, adopt reproducible methods of sampling, collection and measurement. Unnecessary duplication of work within the team should be avoided.

Exact and accurate communication with others is one reason why the modern field geologist makes many quantitative measurements. Another reason is to assist him in communicating with his own memory, to provide a record of his past observations which he can relate to his present work. The trained eye is an excellent judge. Put two near-identical colors or textures side by side and the ability of most geologists to determine which color is lighter or which texture finer is astonishingly well developed. But memory is less accurate than direct comparison. If the two samples are not seen together but in succession, discrimination becomes poorer. If a random succession of other samples intervenes, the accuracy of comparison is largely lost. The field geologist generally works with such a diversity of material that direct comparison, specimen by specimen, is not possible. Instead, a scale is required on which individual observations can be placed by direct measurement, thus creating a consistent order and allowing any pair or all of the observations to be compared.

The field geologist may make many kinds of measurement, generally using a simple tool to enable him to express his observations within a standard frame of reference. With compass and clinometer, he may measure the orientation of strata. With a tape he may measure the thicknesses of individual beds, and with the graticule of a microscope, the size of grains. In the laboratory, banks of sieves may be used to estimate the proportions of size grades of sediment. Geochemical and other equipment may also give quantitative data which can be calibrated against standard samples. Because such data can be quantitative and reproducible, the geologist is in a position to compare his measurements among themselves, and with comparable measurements obtained elsewhere by himself or by others. A large amount of numerical data may thus be at the geologist's disposal, but in a form where the unaided human mind may be unable to grasp its full significance. The computer, in contrast, may be able to extract a great deal of significant information from the data.

1.3. Quantitative and non-quantitative applications; process control; graphic methods

A set of numbers is a much more abstract concept than the collection of objects and properties to which it refers. It is also more generalized, and various mathematical and statistical techniques may be applicable, whether the measurements refer to mud in the Mississippi delta or ostracods in a Cretaceous shale. Since many mathematical procedures are of general application, a range of computer programs for data analysis are widely available.

There are many examples of methods of data analysis, including the following:

Summarization of a set of measurements to give an overview of their general properties and distribution.

Transformation of data to reveal its properties more clearly.

Analysis of variance to compare separate groups of items.

Cluster analysis to suggest groups into which objects can be classified.

Regression analysis to explore the dependence of one property on another or a group of others.

Correlation, principal-component and factor analysis to investigate the relationships between pairs and groups of variables.

Analysis of series to separate and analyze patterns in time and space.

Statistical methods can generally be applied to any number of measurements. Provided data are collected in a consistent manner, therefore, there is no limit to the area or the detail in which comparisons can be made. The ability of the geologist to integrate information derived from many visual patterns is thus supplemented by the computer's ability to detect complex patterns and relationships in long lists of numbers, which the unaided mind could not have fully interpreted.

The application of the computer is by no means confined to numerical data. Some observations are more readily expressed in words than in numbers. Instead of measuring grain size on a quantitative scale, the geologist may divide the range into categories such as very coarse, coarse, medium, fine and very fine grained. If standard examples are at hand, the categories can be consistently assigned by direct comparison. Names are often used, also, to refer to complex concepts, such as reefal facies or glaciated landscape, which depend not on a single measurable property but on an assemblage of characteristics possibly linked by co-occurrence, common cause or common origin.

Words can be handled within the computer as readily as numbers. Relative frequencies can be established, groups of words that tend to occur together can be revealed, names can be arranged in dictionary or other order, specified keywords can be recognized in a length of text. In many geological applications, the ability to handle words, or a combination of words and numbers, is important. An example is the lithological description of a sequence of beds in a measured section from borehole or outcrop. The computer can provide statistics about, for example, total thicknesses of the various lithologies; ratios of lithologies, such as sandstone to shale; facies as determined by frequently occurring combinations of characteristics; or pattern in the sequence of deposition of the beds. The ability of the computer to handle text has given rise to important applications in the management of indexes, inventories and data banks, in the automatic preparation of reports, and in the exchange and communication of information. In many geological projects, the computer data base provides a focal point for sharing data in a consistent, widely usable form.

However, although words can readily be manipulated, the full meaning of free text cannot normally be deciphered by the computer. Perhaps the nearest approach has been in the field of machine translation, where, for example, translation programs in the computer can produce an English version of a scientific paper in Russian. Although some success has been achieved, considerable effort and understanding is usually required from the reader if he is to make sense of a machine translation. Clear, grammatical English is seldom produced in this way. The reason may lie in the deductive reasoning inherent in the mathematics of a computer program. For translation, the ability to analyze language is not enough. Fuller understanding is required, and this may depend on the ability to produce similar language oneself, and thus to have an insight into the thoughts that lie behind the language. It can be convenient, then, to communicate information to the computer in words, but they must be presented in a simple and rigid syntax. Insight and intuition are not to be expected from a computer, and the subtleties and complexities of English prose are well beyond the grasp of present-day machines.

An area in which the computer's contribution is well known is that of monitoring instruments, data capture and process control. Large-scale computer schemes are in use, for example, to control traffic flow in cities or the operation of steel mills. Geological examples are on a humbler scale. A small and inexpensive computer is capable of controlling automatic geochemical or geophysical equipment, such as an X-ray spectrometer or a seismometer, and avoids the need to display and examine large quantities of irrelevant data. One consequence is that much of the quantitative information collected by instruments in a large field investigation may originate as computer data.

It was suggested earlier that one field in which the human being readily surpassed the powers of the computer was in recognizing and analyzing visual patterns. This field is important to the geologist, who frequently thinks in pictures. An interesting development is that of storing graphic data, such as maps and diagrams, within the computer in digital form. Most large computers can generate graphic output on visual display units similar to television screens, or on film, or as large maps or line diagrams drawn on paper or plastic by a pen plotter. Selection, retrieval, and geometric operations such as tilting, rotating, or stretching can be carried out within the computer, and the results displayed graphically. The computer's ability to display, store and manipulate data in two, three or more dimensions is thus allied to the geologist's appreciation of pattern. The combination is a powerful one, and likely to become more so as methods of communication between man and machine continue to improve.

To summarize, it appears that the geologist has many abilities which no machine can emulate. But the computer, like other equipment which the geologist now has at his command, has created new opportunities, and has made new techniques economically feasible. These complement traditional methods, and the combination is leading towards a new view of Earth Science.

1.4. Systems Analysis; systems and models

Although the computer is just one of many machines which are changing the practice of geology, it is more complex than most equipment and more far-reaching in its effects because it deals with a fundamental ingredient of all science, namely, information. The impact of the computer is not restricted to a single specialized area, but is likely at some time to impinge on the work of every geologist. Indeed, the computer has had a major influence on science as a whole as well as on many non-scientific fields. Because its applications are so widespread, a subject exists in its own right to examine areas of human activity to determine where the computer's contribution may lie. It is known as Systems Analysis. The systems analyst is responsible both for making recommendations about new computer applications, and for determining how they should be implemented, for Systems Analysis has a practical as well as a theoretical side. In considering situations in which the computer can assist the geologist, the methods and concepts of the systems analyst may thus be relevant.

The general approach of Systems Analysis agrees with common sense. A typical pattern is:

1. Clarify the objectives.

2. Define the relevant system.

3. Analyze the system into individual components, considering their operation and interaction.

4. Consider the part which the computer can play, and the modifications of the system which might result from computer applications.

5. Design, implement and maintain those aspects of the system which involve the computer.

The technical terms of Systems Analysis, like those of most branches of science, are initially confusing to the outsider, but are essential for clear and exact expression of the ideas they represent. At this stage, two useful concepts of rather wide application may require explanation. They are 'system' and 'model'.

'System' is a concept with wider application than Systems Analysis. A system can be regarded as a collection or set of entities that are interrelated and interacting, and which can sensibly be studied as a whole. For example, a basin of sedimentation could be studied as a system, made up of interacting objects, activities and forces. The various types of sediment entering the basin could be regarded as components of the system, as could the basin floor and its covering of sea and sediment. The rates of subsidence of the basin floor, the debris washed from surrounding mountains, the pattern of currents and tidal activity would all have to be taken into account as part of the relevant system. Such a system would obviously be of interest to a sedimentologist. Three general points about it may be noted here.

First, the components of the system interact with one another in a complex manner. Currents and tidal activity may initially determine the areas of deposition of sediment, and perhaps cause a delta or sand bar to build up. But the deposit in its turn affects the position and strength of the currents. It is not easy, and may be impossible, to disentangle cause and effect. Second, the boundaries of the system are arbitrary. Currents in the sea are included, but the climate and weather conditions which caused them are not. If studied in detail, they would themselves constitute a system at least as complex as that of the sedimentary basin. It is necessary to focus on the system of interest about which information is available. The analyst should be aware of other systems which overlap the chosen area of study, but they must be

B

regarded as external, and of interest only in their interactions with the selected system. The third general point about a system is that it can be studied in varying degrees of detail. For instance, within a basin of sedimentation, the sedimentologist might select a single sand bar for more detailed examination, perhaps on the grounds that it was representative of a particular kind of activity within the basin. His approach to the more restricted system could be the same as that for the basin as a whole. Larger scale effects, such as the distribution of currents in the vicinity of the sand bar, would then be regarded as part of an external system. Again in more detail, he might examine the process of sedimentation in terms of the movement of single sand grains over a ripple on one side of the sand bar. This too would be a valid system for study, with single sand grains and small eddies within a thin layer of water as some of the components.

To summarize, a system is a set of interacting components. It is unlikely to be completely self-contained, and the process of selecting its boundaries may be somewhat arbitrary. There are probably many parallel systems, external to the system of interest, which interact with it to a greater or lesser degree. Any system is likely to be part of a larger system, and may contain many subsystems, each of which might be usefully studied on its own.

The second of the two concepts to be considered at this stage is that of a 'model'. In general terms, a model is a hypothetical construct which represents a system or some aspect of it. As such, it may be a means of providing a formal framework within which an analyst may assemble his ideas about a system. A model may also be designed for prediction or for experiment. It can be important to predict the consequences of modifying a system in circumstances where it is impossible to experiment with the system directly. For instance, it would not be possible, for reasons of size and inaccessibility, to experiment directly with the sedimentary basin mentioned above, nor with the system which constitutes the science of geology. Yet the sedimentologist might wish to predict the consequences of the basin becoming landlocked, or the systems analyst might wish to investigate the probable consequences of introducing computer methods into a geological project. For such purposes it might be possible to construct a predictive model, that is a mathematical or computer representation of the system, which in certain essential ways behaves like the system itself. Experimentation with the model might yield results that could be extended to the system and give insight into the consequences of modifying it. Intuition is important in arriving at a plausible model, but conclusions are drawn entirely by logical deduction from the model. A satisfactory predictive model is fruitful, in the sense of giving rise to predictions that can be tested to extend or to clarify geological theory. It is also plausible, that is, consistent with known fact and established theory, and is no more complex than necessary.

A good model can help in designing an investigation, and considering the role of computer methods in it. Seismic exploration provides an example of an area of Earth Science in which an exact, quantitative model is available. The arrival times of seismic waves can be predicted if the velocities of sound in the layers of the earth's crust are known. The model has been the basis of the design of seismic surveying methods from which accurate results can be obtained, provided that a very detailed analysis of the raw data can be performed. This depends on computer processing. The availability of fast, cheap computing power has led in this case to the successful introduction of more accurate methods.

1.5. The scientific method in geology; the organization of this book

The concepts of the system and the model are useful for an immediate aim, namely, to develop a framework in which to organize ideas about computer applications in geology, that can be followed in the remainder of this book. There are several systems that are relevant, such as the systems of rocks, minerals and geological processes which geologists study. This could be broadened to include the history of geological events which describes the evolution of the earth. Alternatively, one could consider the system of recorded facts and theories which constitute geological knowledge. However, the aim is to assist the geologist who wishes to use computer methods for his work. The system of direct interest therefore comprises the activities on which geologists are engaged.

The science of geology and the workings of the geologists' minds, like the real geological world, are complex and incompletely known systems. The systems analyst, like the geologist, must often work with models that are known to be inexact and incomplete representations of the real world. The so-called scientific method, for example, gives a picture of scientific investigation which few geologists would claim to be an accurate description of their own methods. Regarded as a model, however, it resembles the actual situation closely enough to add to the geologist's understanding of his own activities. In outline, the model might regard the geologist as observing real objects and processes, collecting data, interpreting the data, attempting to explain his observations, and testing his conclusions against the real world.

A distinction is implied between the real world and geological knowledge. It is a distinction between, on the one hand, objects and processes which have an existence quite independent of the scientists who study them, and, on the other hand, the shared ideas of geologists' minds: the maps and descriptions; scientific literature; geological observations and hypotheses. The difference, in short, is between unchanging reality and the scientist's constantly evolving model of it. But this picture is not a factual description of the situation and the implied concept of external reality could be disputed on philosophical grounds. Indeed, the picture itself is put forward as a working model for convenience in developing a few ideas, and can be discarded when it has served its purpose. The system of interest concerns the activities of the geologist rather than of the phenomena which he studies. Even the highly simplified model of the scientific method suggests that the system is too complex for detailed study as a single entity. However, it immediately indicates a possible breakdown into subsystems, each of which can be considered separately. In order to clarify the framework the various subsystems are described in outline below.

The geologist may collect data through his own observation, or by means of an instrument, or may have access to data collected by another geologist. Data are frequently too numerous, too detailed and too unrelated for their significance to be readily grasped by the human mind. Interpretation therefore involves making broader, more generalized statements than the individual observational records. Data are generally regarded as factual, in the sense that another scientist, following the same rules and making the same observations would obtain similar results. Interpretation, on the other hand, is seen as a more subjective activity, in which statements are made that apply to the data, but which can be extended further, to the real world from which the data were obtained. Thus if a number of samples of sand were collected at high- and low-tide levels on a beach, measurements of grain sizes could be obtained. Examination of the data might lead to the conclusion that the sand was coarser at low-tide level than at high-tide mark. This conclusion would be interpretative rather than factual. It is one possible explanation of the observations, but others could be found.

Geological observation is not, of course, a passive activity. The real world is usually observed with particular aims in mind. They may be purely economic, or may involve testing a hypothesis, or developing a particular branch of geological theory. Observation may involve search for a special pattern which the geologist is able to recognize from familiarity with similar examples elsewhere, or from imaginative visualization of the pattern which certain processes might create. A trained geologist automatically observes certain aspects of his raw material with particular care, reflecting a consensus of opinion from past and present workers on what is important and what is not. The process of selective comparison and judgement involved in observation is carried through to the process of recording what has been observed. A petrographer examining a hand specimen of sandstone might casually notice the shape of the fragment, the presence of dust, the color and position of the label. But, these being irrelevant to his examination, he would concentrate more on the size, shape and composition of the constituent grains. He would probably follow a systematic procedure in his description, noting points of significance one by one, probably in about the same order in each specimen described. There are many examples, such as graded bedding and trace fossils, of patterns which have gone undescribed for a long period, simply because their significance was not appreciated. A geologist's observations and descriptions thus constitute a highly personal picture of the real world, seen from a specific viewpoint, colored by the observer's objectives, training and attitudes.

Interpretation is not factual, in the sense that a unique conclusion must result from the application of a set of rules. An interpretation, however, can usually be tested against reality and is thus subject to disproof. It is convenient, for present purposes, to distinguish between interpretation and explanation. Interpretation can be seen as the various procedures which can be applied to data to give broader and more general conclusions. Explanation is seen as arising from a background of geological hypotheses and theory which draws on many sources outside the data, such as the laws of physics and chemistry and knowledge of present-day processes. Hypotheses or models are developed intuitively, rather than by logical deduction, and are based on existing theory which they may refine, modify or extend. They do not necessarily result from new observations, but if fruitful, they can be tested against reality. Explanation can be thought of as finding the correct theoretical background and developing it to account for observations that have been made or data that have been collected.

The pure scientist is likely to be concerned with the development of theory, and his observations and data collection may be aimed at testing and developing hypotheses. The applied scientist or economic geologist may be interested in theory only if it can assist him to refine his search techniques or to improve his extrapolations and predictions. Both are involved in communication, in transmitting the results of their labors to their colleagues, acquiring relevant information from other workers, and relating their own ideas, observations and activities to developments in geology as a whole. Communication can take place by letter or by word of mouth. More formal communication, through the scientific literature, involves not only the primary participants, the author and the reader, but also the editor, referees, publisher, and librarian. Communication of results extends the observational and theoretical basis of geology, thus providing material for new speculation and testing of ideas. It is the starting point for renewed observation and data collection and the beginning of a new cycle of observation, interpretation and explanation. Each of these aspects of a geologist's activities is the subject of a separate section of this book.

HARDWARE

1.6. Problem solving by geologist and by machine

In order to appreciate how the computer can be applied in geology, and to use it successfully, it is necessary to understand something of the way in which a computer works. The abilities of the human mind include observation, reasoning, memory and intuition. They have led, among other things, to devising the computer. The computer, on the other hand, has the possibly more limited ability to accept, store, compare and manipulate sequences of electronic pulses. Man has devised sets of symbols to assist and order his thinking and communication. He uses these in mathematics and for spoken and written language. The symbols can be represented by sequences of pulses, and where the rules for manipulating the symbols can be stated in a simple and formal manner, the computer can be programmed to carry out the required manipulation. A few examples may make this clearer.

The rules for finding the average value of a set of, say, fifty numbers can be expressed in terms of adding each in turn to the current total, and dividing the final total by 50. The procedure is easily broken down into simple arithmetic operations, and can easily be programmed for the computer. The task of arranging a list of words in dictionary order can also be reduced to a sequence of operations of the form: compare the first letter of a pair of adjacent words, and if they are in alphabetical order, go on to the next step; if in reverse alphabetical order, change the sequence of the two words; if they are the same, continue the comparison with the second letter of the words; and so on. Again the basic operations are simple and exact and can readily be programmed for the computer. On the other hand, there are problems which are difficult to program although they would give little trouble to a three-year-old child, such as 'Is the object crossing the road a cat or a dog?' or 'Is the face in front of me familiar or unfamiliar?' The process of solving these problems is difficult to analyze into simple basic operations of addition, comparison and replacement of symbols, and consequently is difficult to program satisfactorily for the computer.

A final example is the problem of preparing a contour map to represent a buried geological surface, penetrated by boreholes at a number of points. The contour lines which a geologist would draw might depend on what he knew about the formation of the surface, the accuracy of the data and the reasons for preparing the map. Without specific information, he might use a generalized procedure of drawing the smoothest possible lines with the most regular spacing possible within the constraints set by the data points themselves. The specific problem of drawing a contour map to represent a surface that resembles, say, a landscape eroded by river valleys and that exactly fits the data points depends on the ability to visualize and judge pattern, and consequently is difficult to program. The more general technique of fitting a smooth surface mechanically to the data, on the other hand, is a successful computer technique. As is often the case, the computer method differs considerably from the manual one, although they have similar objectives. Instead of fitting smooth curves by eye and refining them step by step to achieve a balance of the various criteria, one computer technique is to represent the contoured surface by a mathematical function and use statistical methods to calculate the constants in the function which give the best fit to the data. Since the height of the mathematical surface can be calculated from the function at any point, there is enough information for contour lines to be drawn automatically under computer control.

1.7. Data management and numerical analysis

The examples suggest two purposes for which the computer can be used, namely, manipulation of encoded symbols, and mathematical calculations for data analysis. The rearrangement of words so that they are put into dictionary order is an example of manipulation of data. From the simple ability to compare two symbols, many more complex programs can be created. Large files of data can be searched for items matching a particular description, the appropriate place in a file for inserting new data can be found, old information can be deleted or replaced, or information from several small files can be combined in one. The activities of creating, sorting, merging, editing and up-dating data files, together with information storage and retrieval, constitute the field sometimes known as information or data management.

Information management is based on manipulation, comparison and rearrangement of symbols. Data analysis depends more on conventional mathematical ideas from numerical analysis and statistics. Numbers can be represented in the computer in such a way that the normal arithmetic operation of addition, and hence of subtraction, multiplication and division can be readily performed by the processing unit of the computer. The introduction of mathematical concepts into geology provides a means of generalizing certain ideas and expressing them in an abstract symbolic form. Abstract symbols are of course appropriate for the computer, and generalization of the procedures means that programming work need not start from scratch for every problem. Instead, the aspects of data analysis which are common to several problems can be programmed once and used where required. The geologist can thus call on a wide range of computer applications without having to write his own programs. Much of the power of computer methods lies in the combination of the ability to manipulate and analyze data with the ability to express the procedure in generalized mathematical form.

1.8. Hardware and software

Equipment for computing has come to be known as hardware. In distinction, the programs which are intended to cause the equipment to behave in the desired manner are often referred to as software. The concept of separating hardware and software gives the modern computer its flexibility. Equipment is designed to handle a wide range of software, whereas each program is specific in its application. Many hundreds of programs and data files may be presented to a computer in the course of a day. The same machine may run programs to calculate salaries, correct gravity data and simulate erosion, all within a period of a few minutes. To some extent also, the software can be independent of the hardware, and a program developed on one machine transferred with little change to another. This was not always so. The sequence of operations in some early computers was determined by the wiring, so that a change of program was a time-consuming piece of engineering work. Now, only a few special-purpose control computers have wired-in programs. It is therefore possible to consider programs separately from the equipment on which they are implemented, and programming techniques are left to the next chapter. The machinery and its operation are first described briefly, since some knowledge of both subjects is obviously necessary to appreciate their application in geology.

1.9. The computer functions of recording, storing, processing and communicating

In the introduction (§1.5) it was implied that the activities of geologists involve recording information, remembering, selecting, analyzing and communicating it. A computer system, also, performs these functions. Just as the system of geological investigation can conveniently be considered in terms of the geologist's activities, so the components of a computer system can be considered in terms of their functions. Information must be prepared in a form which the computer can accept. For this, data recording equipment is required. Once in machine-readable form, the information must be made available to the computer by some input device which can translate the data from the form in which they are recorded to the form used internally in the computer. The computer may have to hold the information for some time, on a storage device, before the data are processed by the central processing unit (CPU). The processor must then pass the results back to a storage device or to an output device which can translate them into a form which the user can understand, such as a printed or typed document. The information has to be transmitted from one point to another within the computer, and may have to be communicated to input and output devices at a considerable distance, even thousands of miles, from the CPU. Special data transmission equipment may then be required. The functions of a computer system can thus be grouped into data recording, storage, processing and communication.

1.10. Data preparation devices and methods;
punch cards, paper tape and magnetic tape;
redundancy and error checking

The geologist who becomes involved with computer work generally finds that his first contact with the computer system is in data preparation. Although it is possible to use computers successfully for some time without understanding how they work, some knowledge of data preparation and input and output methods is usually necessary. They are therefore a suitable point to begin a description of the operation of the computer.

The basic method of recording geological information outside the computer context is in written or graphic form, which the human eye, adept at discriminating patterns, can recognize and transmit to the brain. The computer can more appropriately receive its information as a series of electrical pulses. The first step in translating from one form to the other is usually by typing the information on a keyboard, similar to that on a type-writer, connected to a device which can encode the characters into a representation suitable for computer input. It is possible to record information and simultaneously pass it to the central processing unit. For example, messages to control a computer's operation may be entered from a keyboard at the operators console and transmitted to the CPU. To an increasing extent, terminals at sites remote from the computer allow the user to transmit information through data links direct to the computer. Frequently, however, data and programs are prepared away from the computer in a form which can later be read by machine. After they have been checked, edited and corrected, they are added to the sequence of jobs (the 'queue') waiting to be handled by the computer. As the computer can handle information much faster than one operator can type it, data prepared by many operators can be fed into one machine.

Geological records are seldom in a form which a machine can immediately accept. Field notes, maps and sketches, or the results of geochemical analyses, for example, may be difficult for another geologist to understand, and impossible to interpret mechanically. There are two aspects: one is the content and organization of the data, a subject considered in a later chapter; the other is their physical representation. Geological data are often first recorded as words, letters and numbers, as points and lines on a map or chart, or voltage levels in analytical equipment such as a spectrometer. The kind of device used to prepare the data for the computer obviously depends on the type of data. For data represented as letters and numbers, the card punch is a long-established device still in widespread use.

The standard Hollerith or IBM punched card is a thin rectangle of card as illustrated in Fig. 1, on which is marked twelve rows of eighty columns. One character, that is a letter, number, or special symbol, can be coded on one column. Thus one row of data from the coding sheet in Fig. 2 can be recorded on one punched card. The coding is performed on a card punch or keypunch, illustrated in Fig. 3. Blank cards are stacked in a hopper on the right-hand side of the punch, whence they are fed, at the push of a button, to the 'punch station', just below the hopper. From a keyboard, similar to that of a typewriter, the operator can then punch a series of characters or spaces. Each time a key or the space bar is pressed, the card moves forward one column. A small printing device types the character near the top of the card, while an array of magnetically actuated knives at the punching position cut a rectangular hole or holes in any of the twelve rows of that column. The number 8, for example, might be represented by a hole in the eighth row, or the letter C by punching the top row and the third row in the same column. The pattern of holes which represents one character is known as the code for that character, and the set of patterns which represent all the characters is known as the coding system. Several of these are in existence. The Extended Binary Coded Data Interchange Code or EBCDIC Code is probably the coding system which is most widely accepted. The usual set of EBCDIC characters is illustrated on the card in Fig. 1. The letters are all upper-case, thus reducing the number of characters and codes, and simplifying the program by which they can be recognized in the computer.

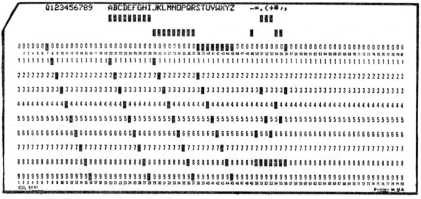

Fig. 1. A punched card, with characters from the ECBDIC character code.

Unit	Thickness	% Conglomerate	% Sandstone	% Siltstone	% Shale	Sedimentary structures present			Number	Basal contact
J-17	47·5	0·0	10·3	43·1	46·6	GB	FL		131	EROS
J-16	32·1	0·0	14·7	28·6	56·7	GB		CB	98	CONF
J-15	03·9	2 3	16·8	42·0	38·9	GB			04	CONF
J-14	56·3	10·7	15·2	38·0	36·1	GB		RM	87	GRAD
J-13	21 3	0·0	16·0	51 3	32·7	GB		RM	53	CONF
J-12	22·0	0·0	17·0	62·5	20·5			RM	48	EROS

```
FORMAT (6X,1A6,5F6.3,3A3,1F6.0,1A6)
GR5XXN  J-12   220    00    170    625    205          RM      48EROS
GR5XXN  J-13   213    00    160    513    327    GB     RM      53CONF
GR5XXN  J-14   563    107    152    380    361    GB     RM      87GRAD
GR5XXN  J-15   039    23    168    420    389    GB             04CONF
GR5XXN  J-16   321    00    147    286    567    GB     CB      98CONF
GR5XXN  J-17   475    000    103    431    466    GB  FL         131EROS
```

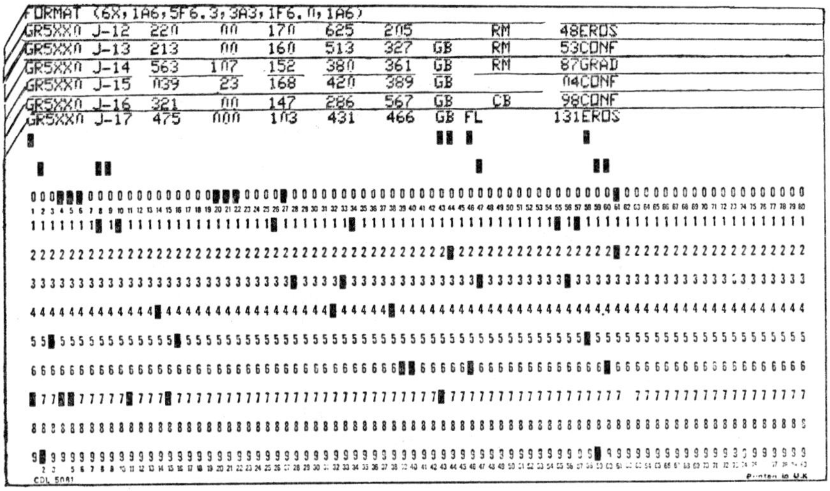

Fig. 2. A table of data, and a set of cards punched from it.

The pattern of holes is difficult for a human being to understand. However, the typing at the head of the card greatly simplifies the problem of interpreting what is punched. The pattern of holes, on the other hand, can readily be read by machine. The device which reads a punched card requires a mechanism to sense whether a hole is present or not. One possibility is a star wheel which rotates a quarter turn when a hole passes over it and remains stationary otherwise, as in Fig. 4. Another possible device for sensing the presence or absence of a hole is a photoelectric cell, which receives light through a hole in the card and is only actuated if a hole is present. A third possibility is an electrical brush which makes contact with a metal plate and completes a circuit if a hole is present. Without a hole, the circuit is incomplete. These methods, and others, are in fact used for machine reading of punched cards. The card is moved, one column at a time, past an array of twelve such sensors, each aligned with one row of the card. The pattern of holes is sensed and transmitted as a sequence of electrical pulses.

On a card punch itself, there is generally a 'reading station'. As a card is punched at the 'punching station' the preceding punched card moves in

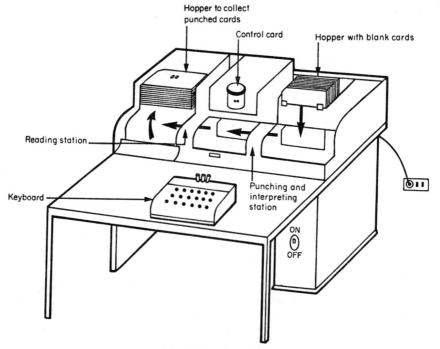

Fig. 3. A card punch.

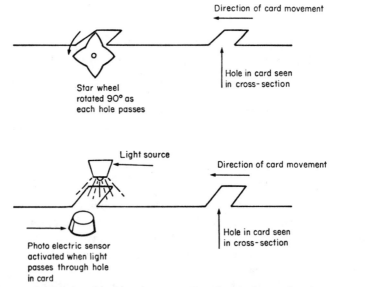

Fig. 4. A star wheel and a photoelectric sensor. Two devices for sensing the presence of holes in a punched card.

synchronization, column by column past the reading station. Signals from the reading station, instead of those from the keyboard, can thus be used to actuate the punch, and information can be duplicated from one card to the following card. A project number could thus be duplicated automatically on to each card of a series. The ability to duplicate also simplifies the procedure for correcting punching errors. For example if 139·7 has been punched instead of 193·7, the part of the card which is correctly punched can be duplicated by pressing the duplication key. When the two columns containing the error are reached, the correct numbers can be keyed in, and the rest of the card automatically duplicated. Most key punches have additional features to assist the operator. For instance, a control card can be prepared for specifying the pattern in which a particular set of data is to be punched. Sets of adjacent columns, known as 'fields', can be duplicated from the preceding card, and it is possible to align fields or skip columns under automatic control. These facilities are provided under computer control in more modern devices. With large data files, prepared for the computer by skilled operators, the cards can be checked and corrected by verification. In this process, the cards are passed through a machine similar to a keypunch, known as a verifier. The cards are read by the verifier while the operator keys in the data a second time, and where the two versions do not agree, the verifier flags the discrepancy, which is then checked and if need be corrected by the operator. If the data were all punched correctly, the operation is redundant, but does give some assurance that the data preparation process, which is particularly prone to error, has been accurate.

In general, similar procedures are used to record characters on paper tape or magnetic tape. Paper tape is seldom more than an inch wide. and at most eight punching positions are normally used. This is reduced in some coding schemes to seven or even five. The internationally accepted ISO (International Standards Organisation) code is illustrated in Fig. 5. It is widely used in various national variants, although many other codes are in existence. Although the ISO code requires only seven channels, an eighth channel is often punched in such a way that the number of holes in every column is even. A mechanical error in which a hole was omitted or an extra hole punched could thus be readily detected as an illegal code, since the number of holes would be odd. The eighth channel is known as the parity channel. It is redundant, since it is not essential for transmitting information, but the redundancy gives a means of finding possible errors. It is thus an error-checking code. The inconvenience of handling paper tape means that it is now seldom used.

Magnetic tape usually has nine tracks, one of which can be the parity channel. Magnetic tape consists of a strong and flexible plastic ribbon with a thin coating of magnetic material. No holes are punched in it, of course, but

Fig. 5. The ISO code for representing characters in a 7-bit binary code.

		b7	0	0	0	0	1	1	1	1
		b6	0	0	1	1	0	0	1	1
		b5	0	1	0	1	0	1	0	1
b4 b3 b2 b1	Row \ Column		0	1	2	3	4	5	6	7
0 0 0 0	0		NUL	DLE	SP	0	@	P		p
0 0 0 1	1		SOH	DC1	!	1	A	Q	a	q
0 0 1 0	2		STX	DC2	"	2	B	R	b	r
0 0 1 1	3		ETX	DC3	#	3	C	S	c	s
0 1 0 0	4		EOT	DC4	⊗	4	D	T	d	t
0 1 0 1	5		ENQ	NAK	%	5	E	U	e	u
0 1 1 0	6		ACK	SYN	&	6	F	V	f	v
0 1 1 1	7		BEL	ETB	'	7	G	W	g	w
1 0 0 0	8		FE0(BS)	CAN	(	8	H	X	h	x
1 0 0 1	9		FE1(HT)	EM	)	9	I	Y	i	y
1 0 1 0	10		FE2(LF)	SUB	*	:	J	Z	j	z
1 0 1 1	11		FE3(VT)	ESC	+	;	K	[	k	{
1 1 0 0	12		FE4(FF)	IS4(FS)	,	<	L	/	l	|
1 1 0 1	13		FE5(CR)	IS3(GS)	-	=	M	]	m	}
1 1 1 0	14		SO	IS2(RS)	.	>	N	<	n	~
1 1 1 1	15		SI	IS1(US)	/	?	O	—	o	DEL

instead small areas of the coating are magnetized in one of two directions by a writing head, which contains an electromagnet actuated by electrical pulses as the tape is wound past (see Fig. 6). Reading heads can sense the direction of magnetization by induction effects. Recording densities of 1600 characters per inch and more are usual on magnetic tape which thus contains far more information in a small space than either punched cards or paper tape. Partly because of the high density of information, tape can be written or read at high speed. The recording and sensing, or reading and writing devices are correspondingly more expensive, and at present only intensive usage justifies the high cost of full-size magnetic tape units for data recording.

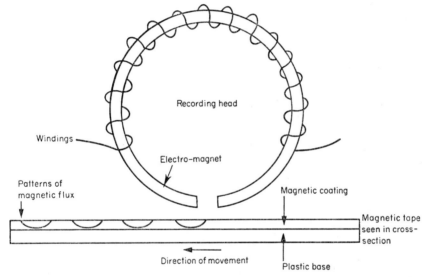

Fig. 6. The writing head on a magnetic tape deck. Pulses of electricity passing through the windings on the recording head create patterns of magnetization on the tape or disk. In a similar arrangement, the patterns are read when they create magnetic flux in the reading head, inducing voltages in the windings.

1.11. Balance, compatibility and trade-off in data recording; key to disk or cassette tape; direct data entry

A single keyboard attached to a fast magnetic tape unit is an example of an unbalanced system. The total resources of the system cannot be used because it can operate only at the speed of its slowest component. Similar problems arise with computer systems. For example, a computer would be delayed if all information had to be entered through a single keyboard. Since slow input would be the cause of its slow operation, the computer could be said to be input-limited or input-bound. The addition of a fast card reader might clear the input bottleneck. If the results of all the computations were printed on an electric typewriter, the system might then be output-bound. A faster output device, a line-printer, could be added with the possible result that the processor would be unable to keep pace with the reading and printing speeds. The system would be processor-bound. The ideal of a perfectly balanced system is unlikely ever to be achieved, but it is at least possible to ensure that the more expensive components are used to the maximum extent. The same general principles apply to any system, whether or not a computer is involved.

A possible solution to the lack of balance in the keyboard to magnetic-tape system is to connect several keyboards to one tape unit. A number of operators could then pass information simultaneously to one recording device. In practice, a small computer might be used to ensure that information from the different operators is placed in separate files, and a magnetic storage device such as a disk or drum, described later, might be used in place of magnetic tape. Another possible way of overcoming the lack of balance in the system is to use a slower and cheaper tape recorder. In place of the half-inch wide tape which is the usual standard for computer use, tape cassettes similar to those developed for portable audio tape recorders, or 'floppy disks' (diskettes) about the size of a 'single' record, could be used. Designed to work at comparatively slow speeds, the recording and reading devices are correspondingly cheap. However, although most medium to large-sized computer systems can accept input from punched cards or half-inch magnetic tape, few can accept small tape cassettes. The inability of one part of the system, such as the computer, to accept output from another part of the system, such as the tape-cassette recorder, is generally referred to as incompatibility. In this case, it can be overcome, at additional expense, by the use of a separate device to convert the output of the tape encoder to a form suitable for computer input. Compatibility, or the lack of it, is an important consideration in systems generally, and examples will be given later in other contexts.

If the attributes of the various data recording systems are considered, it can be seen that advantages in one aspect are often gained at the expense of another aspect. For example, the convenience of use of a card punch is offset by the relative cheapness of punch-tape equipment. The need to consider the trade-off between two or more desirable properties is again a general feature of the analysis of systems, and other examples will be given later. Clearly, the optimum system is one in which the characteristics correspond most closely with the user's objectives and requirements.

The data entered on a cassette or diskette recorder is simultaneously typed on paper or displayed on a visual display unit. The data can be checked and errors corrected before the information is passed to the computer. It may be possible to transmit the data by telephone line or direct wire to the computer from the recorder, which is then known as a 'terminal'. Local storage is not then essential, the alternative being direct data entry from the keyboard to the computer. Although it places an additional load on the computer, this can be an economical procedure for small amounts of data or programs, particularly if rapid processing is required.

1.12. Data capture at source

Data recording has been considered so far as a two-stage or multi-stage process. The geologist, for example, may make notes in the field and later extract some information from them for computer processing. The selected information might then be copied to coding sheets and either recorded by the geologist or given to a keypunch operator. It would not be unusual for the data to be typed at least once before being prepared for the computer. Since the mechanical processes of transcribing and keying the data are being repeated several times, the system contains an element of redundancy, and there may be an opportunity to make it more efficient. For example, if the data were keyed initially on a cassette recorder, the typewritten copy would be available for checking and as a byproduct, the tape can be used for computer input. Corrections to the data can be made by editing the tape, and retyping the correct portion is unnecessary. Similarly, selections of data can be copied from the tape for computer analysis without the risk of introducing new errors during retyping.

Provided the facilities are available, it may be possible to design a system for data collection and recording in such a way that the data are prepared in machine-readable form at the earliest possible stage. Various kinds of output can then be derived from the same basic record. Data preparation not infrequently accounts for half the cost of a computer study, and savings of this kind may be worth attempting. There is a danger, however, that unless the eventual purposes for which the data will be required are known in advance, much time and effort can be wasted in recording data that are so complex that they are useless for any practical purpose. Where several procedures require the same data in different formats, the most complex format can be considered first and will normally determine the appropriate manner of recording the data. For example, if a lithological description is required for publication in full upper and lower case together with punctuation, it may be possible to record it initially in this form and subsequently use it also for input to a computer program where only the words, considered as sequences of letters regardless of case, are significant. If upper and lower case letters are compared in the code in Fig. 5, it can be seen that merely by dropping the sixth channel on input, all letters can be changed to upper case. In contrast to the ease with which information can be degraded, it is often exceedingly difficult to increase the information in a record. The process of adding

c

punctuation to a sequence of words, for instance, is by no means mechanical, and probably requires an understanding of the meaning which only a geologist can provide.

Needless repetition of the same activity can be reduced further if it is possible to record the data initially in machine-readable form, a process sometimes known as data capture at source. This technique is particularly effective where the information is provided by automatic equipment such as data loggers attached to spectrometers for geochemical analysis or geophysical sensors such as gravity meters or seismometers. If the original signal is in a form which varies continuously, such as a voltage level, this analog signal can be sampled at predetermined intervals and changed by means of an analog-to-digital converter to a sequence of digital pulses for input to a tape recorder. Thus the output of data from the analytical equipment could be in a form which could be used as computer input. Cassette tape is suitable for many laboratory applications, since the data are normally sequential and the output rates are not high. Data loggers are available that can operate in extreme conditions, such as being left exposed for long periods to log say, weather or stream flow conditions, or being used on board ship to record ocean temperature. If data rates are high, or immediate processing of the data is required, perhaps to determine the next step in an analytical process, the data may be transmitted direct from an analog-to-digital converter to a small computer.

Where data are obtained by human observation rather than by instruments, source data capture is still possible. One technique is the mark-sense card or sheet which is printed with a design to indicate positions which have a special meaning in terms of the data. A geologist might, for instance, record results obtained by examining rocks under the microscope by marking appropriate boxes in a form. When passed through a mark-sense reader, pencil marks in the boxes are sensed by an array of photoelectric cells and the information transmitted direct to a computer or to an output device such as a magnetic tape encoder. Mark-sense cards are appropriate for surveys in which information is recorded for only a limited number of predetermined categories. It eliminates the keying operation and the inevitable associated errors by ensuring that the initial data collection is in machine-readable form. With complex data, however, the value of the mark-sense sheet is doubtful.

The so-called handwriting reader gives rather more scope for recording complex information. It is capable of recognizing a limited number of symbols, such as letters and digits, if these are carefully written in set positions on a sheet. Simple stratigraphic data have been recorded successfully by this means (see Piper, 1971 *et al*). Recognition of a full character set in any position on a page is possible under certain conditions. If a suitable typeface is used, for instance, type-written documents can be read under favorable conditions by mechanical equipment. This does not, however, eliminate the keying operation, and the technique is not widely used in geology.

1.13. Digitizing graphic data

The methods mentioned above are concerned with the preparation for the computer of information which was originally in the form of letters, numbers or electrical values, such as voltages in geophysical recorders. The geologist may also wish to extract information from maps, seismic sections, charts and diagrams in order to analyze it by computer. In order to make graphic information available for computer processing, it is generally necessary to give the location of each point in digital form, that is, as a pair of rectangular coordinates (x, y), giving the distances of the point from two reference lines at right angles. The first coordinate, x, might for instance be the easting of a point and y its northing, or x and y might simply be the distance in inches from the left and lower edges of the map, as in Fig. 7. To obtain manually the coordinates of a set of points on a map, a simple procedure is to obtain or prepare a transparent overlay on which a rectangular grid is inscribed. The overlay is placed precisely on the map so that the corners of the map and the grid coincide, and the two are taped firmly together. Coordinates of each point of interest can then be read and listed for recording later on cards or tape. Straight lines, such as linears on an air photograph, can be represented

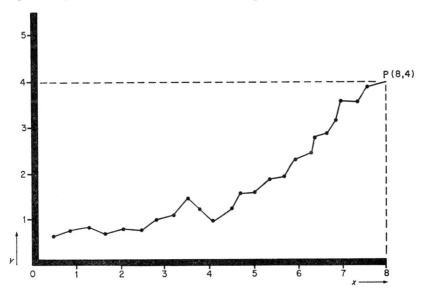

Fig. 7. The coordinates (x, y) of a point. The point P has coordinates $(8, 4)$. The original smooth line is represented approximately by a set of straight line segments.

by two points each, one at either end of the line. A curved or irregular line, such as a geological boundary, is more difficult to digitize. It is possible to regard an irregular line as consisting, as a first approximation, of a series of short straight-line segments, as in Fig. 7, and to represent it as a series of coordinates, each pair representing the end of one segment and the beginning of the next. The accuracy of the representation depends on the length of the segments.

Digitizing the information on a geological map or horizons on a seismic section is a tedious process, and equipment has been devised to enable much of the work to be done automatically. The automatic digitizer has various forms and various names, such as curve reader or pencil follower. Typically, it consists of a digitizing table, to which the chart or map is attached, a pencil or stylus which is placed on the points to be digitized, and an output device, such as a tape unit, on which coordinates of the position of the stylus are recorded in machine-readable form. A number of methods are used for sensing the position of the stylus and translating it into digital coordinates, including electrical and magnetic detecting devices.

From the point of view of the user, the typical procedure for using a digitizer is approximately as follows. First, the map is taped in position with its edges parallel to the edges of the table. Four reference points at the extreme corners of the map or at grid line intersections are digitized first to locate the map, and their map coordinates entered from a keyboard. The stylus is then placed in sequence at each point that is to be digitized and a switch, perhaps in the form of a foot pedal, is pressed to record the measurement. If an irregular line is to be digitized, the usual procedure is to move the stylus steadily along the line while a time switch causes the location of the stylus to be recorded at equal time intervals. Since the tracer is moved more slowly where the line is irregular, the coordinates are more closely spaced there, with the desirable result that the intricate sections are recorded in most detail.

Less manual intervention is required in some forms of digitization, in which, for example, a photoelectric eye locks on to a line and follows it, digitizing points at preselected intervals. Although appropriate for recording simple curves or lines, complete automation is not justified in many geological applications. The 'lock-on' method requires human intervention to label the lines and indicate the correct route where lines meet or cross. This can be less accurate and more laborious than the semi-manual procedure of moving a stylus.

Some useful additional features are available on some digitizers. In some models, a photograph can be projected on to the digitizing table instead of attaching a diagram to it. In others, a scriber is used instead of a stylus, marking the digitized lines on a plastic overlay. Errors can thus be spotted immediately, and the risk of repeating or omitting a line is reduced. A keyboard may be linked to the output device, so that lines can be identified by labels during digitization, and descriptive information can be included in the digital record.

Much graphic information in geology is in the form of lines or points, appropriate for digitizing as described above. Some graphic information, however, such as air photographs, seismic records or thin sections of rocks, consists of continuous patterns of varying density or color. Such patterns can be represented digitally by measuring the density of the image at a set of grid points. The density around each point can be measured photoelectrically and recorded as a number between, say, 0 and 64, adjusted so that 0 represents pure white, 64 black, and intermediate numbers progressively darker shades of grey. The image can be recreated as on a newspaper photograph by plotting dots with diameters dependent on the size of the numbers. Analyzing the information by computer is not easy, because of the large number of data points, but the method is successfully applied in image enhancement and pattern recognition for military purposes and for interpretation of air and satellite photographs.

1.14. Representation of data within the computer; tapes, disks and main memory

Data may be prepared for the computer on ancillary or auxiliary equipment, that is, equipment which is not directly connected to the computer. Devices of this kind are said to be off-line, in contrast to on-line devices which are connected by electrical circuits to the processing unit of the computer. On-line devices which are not part of the mainframe are known as peripheral equipment. Input peripherals, for example, accept information in computer-readable form, such as punched cards, and transform it to electrical pulses, which are passed to the processor unit.

The ancillary equipment for data preparation has been described in detail, because the geologist is likely to come into direct contact with it. The peripheral and mainframe equipment, on the other hand, is normally operated by a specialist computer group. The user is not required to understand the operation of the computer, but simply to ensure that his work is prepared in a suitable form for the receptionist or machine operator to accept for presentation to the computer. Results of the computation are later returned to the user. Although geologists thus tend to be insulated from the computer operations, a knowledge of the general principles of operation is desirable for effective and efficient programming. At the most fundamental level a computer must have the ability to store and to operate on information. In the conventional digital computer, both functions depend on various kinds of bistable state device, that is, a device which can exist in either one of two states, can be moved from one state to the other and remains in that state until actively altered. An example is the common light switch, which has two states, on and off. Each punching position on a punch card or paper tape can either be in a punched state or a not-punched state, but can be altered in one direction only. Magnetic tape, however, in which each recording position is magnetized in one of two possible directions, can be altered either way. Surfaces which, like tape, have a magnetic coating, are a principal means of information storage in computers.

Tapes are convenient for storing data which are arranged and required in a constant sequence, because the information is recorded or read sequentially as the tape passes the writing and reading heads while being wound from one spool to another, see Fig. 8. They have the disadvantage that if a specific record is required, all the intervening information must first be read. Other

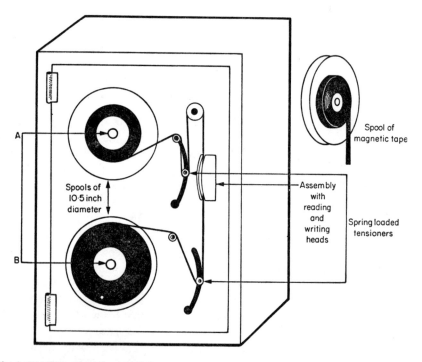

Fig. 8. Small magnetic tape deck for minicomputer. A normal spool can hold 2400 feet of half-inch wide tape. The spindles A and B rotate, winding the tape from one reel to the other, past the reading and writing heads. Magnetic tape and take up spool are loaded on the spindles in the positions shown. The operator threads the tape as shown as part of the loading process. On more expensive tape decks, threading is automatic.

mechanical arrangements are possible which enable the head to reach the relevant information without passing all the intervening records. This is known as direct or random access, as opposed to sequential access on magnetic tape. One method is to carry the magnetic coating on a disk or drum rather than on tape. The disk resembles an LP record in size and shape, and is rotated on a spindle. Information is stored on concentric tracks on the disk and read or recorded by a movable head or a series of fixed heads. The moving head is moved mechanically to the desired track, whereas fixed heads are positioned one on each track, and the appropriate head is activated by a switching mechanism. Because electronic switching is used rather than mechanical movement, the fixed head disk generally allows faster access to data than a moving head disk unit. The latter, however, can have retractable heads, and consequently it may be possible to exchange disks on one disk drive. Thus, several disks can be stored in a cabinet, and, like magnetic tape, when a specific set of information is required, the relevant disk can be removed manually from the cabinet and loaded on the drive. Drums are less commonly used. They are similar in principle to the fixed-head disk, but the magnetic coating is carried on the outside of a rotating drum. A large disk can typically store about a hundred million characters of data, and a tape considerably less. The relative cost of storing information increases from tape to exchangeable disk to fixed disk. The time to access specific items of information increases as the cost decreases. There is thus a trade-off between costs of storage and access time.

Access to program instructions and the primary data used in the course of a calculation must be very rapid in order to keep pace with the speed at which the central processing unit handles the instructions. For even more direct and rapid access than from disk, therefore, storage in the main store, also known as fast access memory, of the computer is generally on magnetic cores, or, on more modern machines, on semiconductors. A core is a tiny ring of metal alloy which can be magnetized in one of two directions, clockwise or counter-clockwise, and can thus act as a bistable state device. Its function is analogous to the recording position on card, tape or disk. But, whereas physical movement is involved in positioning a reading head at the relevant item of information, the reading and recording mechanism is an inbuilt part of core store. Reading and recording wires pass through the center of each core. A current passed through the recording wire determines the direction of magnetization, depending on whether it flows in the positive or negative direction. The reading wire carries a smaller current, too small to alter the direction of magnetization, but large enough to enable the existing state to be detected by induction effects. Rapid electronic switching makes it possible to reach specified positions in core without time-consuming mechanical movement.

1.15. Bits, BCD and binary arithmetic

A single core, like the other bistable state devices mentioned above, carries only one binary unit of information, that is, the amount of information contained in making one of two choices, such as yes or no, on or off. The two choices can be represented numerically as 0 and 1. The amount of fast access memory in a computer can be measured in binary units, often referred to as bits. It ranges from perhaps 128 000 on a small computer to eight millions or more on a large computer. In practice, as mentioned below, the bit is an inconveniently small unit for measuring storage capacity. The binary unit is the basis of a simple counting scale, the binary scale, which consists essentially of counting in twos. Because man has ten fingers, we are more accustomed to counting in tens, and writing in successive powers of ten. Thus, the number 1984 means $4 \times 10^0 + 8 \times 10^1 + 9 \times 10^2 + 1 \times 10^3$. In a similar way, binary numbers are represented by successive powers of two. Thus 1001 in binary means $1 \times 2^0 + 0 \times 2^1 + 0 \times 2^2 + 1 \times 2^3$. The value 1984 in binary is 11111000000. In this way, numbers can be stored on bistable state devices and represented in the computer in binary form. The number written above could be held in a sequence of 11 bits, by 11 bistable state devices.

In the earlier description of punched cards and paper tape, the use of a binary code to represent letters or numbers was mentioned. Magnetic tape with nine channels can hold nine bits of information in one strip. In Fig. 5, a conventional coding scheme for characters is shown. Using this code, the characters 1984 would be represented as 0110001 0111001 0111000 0110100. This is known as binary coded decimal or BCD, since each digit is coded separately, as opposed to the binary representation. Binary coded decimal is appropriate for information which is to be held and manipulated as characters. However, if calculations are to be carried out on the numbers, they must first be translated to binary form.

The function of the reading and recording wires in core store, or the reading and writing heads on disks or magnetic tape units, is to convert bits of information stored in magnetic form to pulses of electricity and vice versa. The nature of the information does not change, but only its physical representation. For instance, an electrical signal more negative than -3 volts might represent 1, and above 3 V represent 0. The correspondence between pulses of electricity and bits of information makes it possible to transmit information between computers and between different parts of the same computer.

Mathematical operations within the computer are carried out on electrical pulses, each of which represents a bit of information. Mathematical calculation in binary arithmetic is straightforward. For instance, $1 + 1 = 10; 0 + 1 = 1; 0 + 0 = 0; 1101 + 0110 = 10011$. Where binary digits are represented by pulses of electricity, it is possible to design electrical circuits to perform this kind of arithmetic, as shown in Fig. 9. Such circuits require rapid switching devices that can be controlled electrically, and the semiconductor device or transistor is well suited to this purpose. A switch with two stable states, can, as indicated above, act as a storage device for one binary digit. A transistor can act as a switch that can itself be actuated by an electrical impulse. The impulse could be a bit of information resulting from an earlier step in the calculation. Thus, the results of one step in a calculation can determine the next step. For instance, two binary digits could be compared by subtracting one from the other. If both are the same, the result will be 0, if different, the result will be 1, or -1. The result could be used to set a switch and hence control the path and destination of the next electrical pulse.

1.16. Bytes and words; integer and floating point;
registers and addresses

Simple basic operations can thus be built up to perform such operations as subtracting, multiplying and dividing binary numbers; of transferring information from one point in the computer to another; comparing values of two numbers; controlling the flow of information on the basis of stored instructions, or the results of calculation or comparison. Although the binary digit is ideal for these purposes, it is inconveniently small for the human being who plans or programs the sequence of calculations. It is therefore usual to group the binary digits as bytes, groups of eight bits, or as

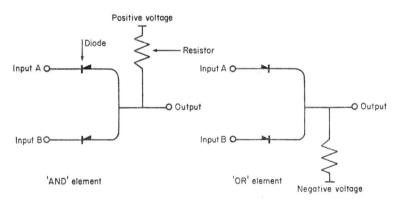

Fig. 9. Simple circuits to perform binary operations. A signal, in the form of an appropriate positive voltage, arrives at one or both of the inputs. The output voltage in the 'and' element will be equal to the less positive input, and in the 'or' element to the more positive input. If the positive input signal represents 1 on the binary scale, then the output from the 'and' element will be 1 if A is 1 *and* B is 1. The output from the 'or' element is 1 if A is 1 *or* B is 1.

words, commonly of 16 or 32 bits. Thus a binary-coded decimal character can be stored in one byte, and either four BCD characters or a binary-coded number in a 32-bit word. The word is treated as a unit and operations performed on all bits of the word simultaneously. The programmer thus thinks of adding, comparing or transferring words rather than bits. The representation as a word of a binary coded integer number has been described, but most quantitative values in science are measured to a decimal fraction rather than as integer numbers. They cannot therefore be stored in the binary form previously described. Instead, a value, say 863·4, could be represented in the form $0·8634 \times 10^3$, with the two numbers 8634 and 3 being stored, together with a sign bit to indicate whether the value is positive or negative. In a computer, working in binary arithmetic, a base of 8 or 16 is more convenient than 10. This is known as floating point representation, as opposed to integer representation. It can handle large and small numbers generated during a calculation, to an accuracy of about six decimal digits, but if this is inadequate, double precision can be used, in which a value is stored in two computer words.

Registers, which are special storage areas in the CPU that hold one complete word, are used to hold the current value during a calculation, and keep track of the sequence of events in a program. The accumulator registers in the arithmetic unit hold results at each step in a calculation, to use in the next step or for transfer to main memory. The instruction register holds the coded instruction which describes the current operation, and other registers hold information about the next item of data and the next program instruction. Data generated or required during the calculation, and the set of program instructions, are held in fast access memory. Each word or location in memory has an address, a number specifying its position. Program instructions are held at a sequence of consecutive addresses. As each instruction is obeyed, one is added to the appropriate register, which thus contains the address of the next instruction. This address is read by the control unit, which automatically transfers the appropriate word to the instruction register, and the circuits of the arithmetic unit carry out the operation indicated by the instruction, returning the results to the accumulator. From this simple basis, methods have been devised which enable the geologist to communicate with the computer in a language designed for the user's convenience, and not unlike the familiar language of mathematics.

1.17. Output devices and methods; the line printer, graph plotter and other devices

Input methods, by which the geologist passes information to the computer, have already been described. The use of punched cards and paper tape was mentioned in this context. When a computation is complete, the converse operation is required of making the results available in a form which the user can understand. This operation, and the results themselves, are known as computer output. Binary numbers are obviously not a suitable way to present output. The results are therefore translated into binary-coded characters in the computer, and transmitted as electrical signals to the output peripheral, where they control a mechanical activity, such as the operation of an electrical typewriter.

There are output devices analogous to most of the input peripherals mentioned earlier. Corresponding to punched card and paper tape readers are card and tape punches. The computer can 'write' information to magnetic tape or disk as well as reading information from them. These forms of output cannot be read conveniently by the user. They may, however, be required for longer-term storage and later input to the same or another computer, or for input to a slower ancillary device which operates off-line. Corresponding to the typewriter keyboard used for entry of information to the computer is the output typewriter. This is a comparatively slow device, typing at some 10 to 30 characters per second, and it cannot keep pace with the output from a large computer. Much faster printing devices have therefore been developed.

The most widely used output device is the line printer. A fast line printer can print over 1000 lines per minute on fan-folded paper, generally with up to 132 characters on a line. The character set is restricted, for greater speed and efficiency, and it is quite usual for only upper-case letters, numbers and a few special characters to be available. A more extensive print chain with lower-case letters and other additional characters is optionally available on some printers, the wider choice being gained at the expense of speed.

Most line printers operate entirely by mechanical means. A chain of print symbols is moved past a row of hammers, one for each print position. Appropriately timed movements of the hammers cause the correct print slugs on the chain to strike an inked ribbon against the paper. Alternatively, a set of continuously rotating thin rollers arranged as a barrel or cylinder with one roller at each print position, each carrying the full available character set, may be individually struck against the ribbon and paper when the appropriate character is in position to print a complete line during each rotation. Fast printing devices operating on other principles have been introduced in an attempt to lower costs and improve reliability by reducing the mechanical complexity. One such device of particular interrest to the geologist is the dot raster printer. Characters are built up from a pattern of dots produced by thermal or electrostatic means on sensitized paper, or by ink jets which throw small dots of liquid ink on the paper. Since each potential dot position is fixed, the styli which produce the dots do not themselves move, and the characters are determined by the electronic selection of the pattern of styli that are activated. The dots can not only form characters, but on some equipment can be activated individually by the computer to produce diagrams, (see Fig. 10).

The digitizer was mentioned previously as an input device which can be used to prepare graphic data for the computer. The analogous output device is the graph plotter. With the digitizer, lines on a diagram are followed by a tracer which automatically passes information to the computer about its present position in x and y coordinates. With the graph plotter, information about successive (x, y) coordinates is passed from the computer to the plotter. Commands from the computer cause the pen to move to the correct (x, y) location, to be raised for movement between lines, and lowered to the paper where a line is required. In flat-bed plotters, paper is fixed to the plotting table. Lines can thus be added to existing base maps. For accurate work, where possible shrinkage or distortion is important, a plastic stable base material can be used instead of paper. Ball-point or liquid ink pens of various colors and line-widths can be used. Some plotters have a pen carriage with two or more pens which can be selected by the computer program. The plot can therefore be in more than one color. Plots may be from one to several feet square, to an accuracy of 1 mm to 0·1 mm or better, depending on the plotter. Drum plotters plot on a roll of paper, typically 3 ft wide and 120 ft long. The pen moves on one axis only, movement in the other direction being achieved by rotating the paper over the drum from one roll to another.

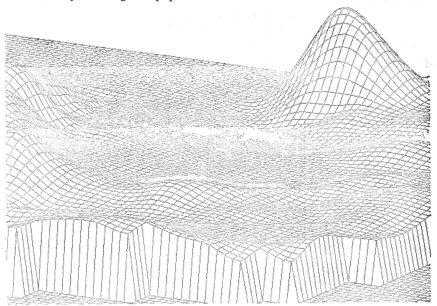

Fig. 10. Part of a diagram prepared on a dot raster plotter.

The drum plotter is generally cheaper and faster, but less accurate, than a flat-bed. Plotters can be operated on-line to a computer, but since they are comparatively slow devices, off-line usage can be preferable. A magnetic tape is prepared on the computer with coded instructions for step by step movement of the plotter, and this is later transferred to a separate tape deck from which the plotter is controlled.

Line printers and plotters are available at most computer installations. Less widely available are devices such as typesetters which produce high quality print for publication, and microfilm recorders, which are perhaps the cheapest and fastest way of obtaining very large quantities of output information. The microfilm is produced by photographing a cathode-ray tube similar to that in a television set, on which characters, symbols and diagrams are displayed. Visual display units (VDU's), on the other hand, are cathode-ray tubes designed for direct observation rather than for photography. Although in some cases a permanent record, known as hard copy, can be obtained by a photographic process, the main use is for ephemeral displays which can be discarded as soon as they have been examined.

1.18. Multiprogramming, the buffer and the queue

On most computer installations, several jobs can be processed simultaneously. Using a technique known as multiprogramming, several programs can be held in memory at one time. Since the jobs have differing requirements for peripheral devices, a supervisory program, known as the executive or operating system, is used to select a job-mix which maintains a balance within the computing system by ensuring that work is not delayed unnecessarily by uneven demand on peripherals. An advantage of punched cards as an input device is that they can act as a buffer, in which information is accumulated character by character at the speed of the operator, and read card by card at the speed of the computer. Card decks prepared for various jobs on different key punches can be stacked one behind the other as a queue in the card reader. These concepts of the buffer and the queue are relevant to many systems, including geological ones, but are significant here for providing the ability to service many peripherals efficiently. An example is provided by a computer system which is designed to service a large number of remote terminals, some of which might be keyboards on which programmers were entering information at a slow speed. A random access storage device, such as a disk, can be used to assemble information as it is presented to the computer, organizing it as a queue of program segments and data sets, which the processor can handle in sequence. The processor unit does not then stand idle while incomplete information is received, but instead processes only complete job segments. The queue ensures that the various items of incoming and outgoing information can be assembled in a meaningful manner, and stored for access in the correct sequence of priorities for the next stage in processing or output.

1.19. Remote access; the minicomputer; the desk calculator

Operating systems which can handle multiprogramming and data transmission have led to the development of data communications, and networks of computers connected by data links. A single large computer, which may have economies of scale compared to a group of small machines of comparable total power, can serve many dispersed users equipped with their own input and output devices at points remote from the computer. The telephone network is generally used to connect the remote users. The remote terminals may be as simple as a keyboard and a typewriter or VDU, possibly with local storage on cassette or floppy disk, or may be a small computer equipped with a full set of its own peripherals. An important result has been that not only can some functions be centralized, but others can be decentralized, by making the minicomputer more effective. The minicomputer is a relatively small, relatively simple computer, which, by avoiding some of the overheads associated with a large complex installation, is comparatively cheap, and good value for money in performing simple or specialized tasks. Data links make it possible to take advantage of its cost-effectiveness for the work that it is designed to handle, while giving the user access to the full range of facilities on the large remote computers to which the minicomputer is linked. The effects of availability of transmission and data links are considered in the chapter on communication.

The abilities and cost of the minicomputer overlap with those of the desk calculator, and the distinction between them is somewhat arbitrary. Extensive programs can be held for the calculator on magnetic cards or cassettes, and a range of peripherals including printers, digitizers and plotters, is available. Programs are entered initially from the keyboard, on which keys are available to represent mathematical operators such as multiply and subtract, and functions such as square root or sine. The program can generally be stored for re-use if required. The calculator, however, lacks the versatility of the computer. The operating system is limited, and no choice of language is available. Processing is restricted to numerical data. Widespread exchange of programs is not possible. The peripheral devices are generally more limited and slower than those on a computer. However, for numerical calculations which are not too complex, where keyboard entry of data is convenient and rapid response is required, and particularly where access to a computer is difficult, the desk calculator may offer a desirable and cost effective alternative to the computer.

SOFTWARE

1.20. The interface between hardware and software;
machine language

The equipment which makes up a computer configuration, such as the mainframe with main memory, processor, registers and interfaces, the peripherals such as magnetic tape and disk drives, card and paper tape readers and punches, line printers and plotters, together constitute the computer hardware. In contrast, the programs which control the operation of the computer are sometimes referred to as software. At the most basic level, of course, the manner in which a computer operates is controlled by the arrangement of its electrical circuits and electronic devices. In a general purpose computer, this aspect of the computer's design is hidden from the normal user. At the basic level, patterns of binary digits are stored in sequence in store, and each bit pattern in turn is converted to electrical pulses which may be modified by the logic circuits of the computer or may themselves set switches which modify the circuit. At this level, the binary coded patterns are commands in machine language, which control such operations as copying a word held at a particular location in store into the accumulator, adding a number held in a storage location to the quantity stored in the accumulator, transferring a number in the accumulator to a given location in store, transferring the contents of the output buffer to the line printer or the contents of a location in main memory to disk. There are typically a fairly small number of these basic commands.

Machine language instructions can be specified by a few bits at the beginning of a word, the rest of the word being used for the address, if any, to which the word refers. Such operations are very rapid, the time being measured in microseconds, that is, millionths of a second. They are not suitable, however, for most programmers to employ directly. Firstly, few users wish to describe calculations at this basic level. Secondly, the instructions are difficult to memorize, having no mnemonic value. Thirdly, they reflect closely the design of the computer, hence a machine language program cannot easily be transferred from one machine to another.

1.21. User- and problem-oriented languages;
assemblers and compilers;
compatibility and standards

To make the programmer's task simpler, a range of user-oriented and problem-oriented languages have been developed. Special programs translate instructions in these languages into sets of statements in machine language.

The user-oriented languages closest to machine language are known as low-level languages. An example is ASSEMBLER language, which is a simple mnemonic code similar to machine language. Low-level languages are largely specific to one type of computer. High-level languages, on the other hand can be largely independent of the computer on which they are implemented. They allow the user to express his instructions to the computer in terms not unlike those of mathematics. The process of translating a high-level language program into machine language is known as compilation, and the program which carries out the translation is known as the compiler. There are a number of languages, notably FORTRAN, PL-1, ALGOL and BASIC, which are widely used for scientific programming, and compilers for at least one of these are available on most computers apart from small or special-purpose machines. It is possible, at least in theory, to implement the same FORTRAN program on many different computers. The program is thus said to be portable, and compatible with many computers. In terms of the earlier definition of 'compatible', the output from one part of the system, namely the programmer, is acceptable as input to another part of the system, namely the computers.

Compatibility can thus be achieved at the interface between two parts of the system. By defining precise standards at the interface, communication is simplified without the need to standardize other aspects of the system, such as compilers or computers. As in the case of character codes, international agreement has been reached through the International Standards Organisation and the corresponding national organisations, such as the American National Standards Institute (ANSI), on the precise form of the STANDARD FORTRAN language. Many manufacturers have prepared compilers which include extensions to STANDARD FORTRAN or additions to the language. If the programmer makes use of these features, he should be aware that he is increasing the difficulty of running his program on another machine. Conversely, there is no guarantee that a program will work simply because it has been used successfully on another computer.

1.22. Programs, subroutines, instructions

The geologist who wishes to write or modify a program, rather than using an existing one, is most likely to prepare programs in a high-level language, and the properties of such languages are therefore worth considering in more detail. A program is a set of instructions which cause the computer to behave in a particular way, generally to accept data, carry out various operations and mathematical calculations on the data, and to return the results to the user. For example, a program might be written to compute descriptive statistics for results of mechanical analyses of present-day sediments. The raw data, consisting, say, of the weights of sediment retained in each of a set of sieves, might be punched on cards. The results of running the program might be a listing of the raw data, followed by a table showing the mean, standard deviation, skewness and kurtosis calculated for each sample.

A program, such as that described above, consists of a number of steps. The raw data are read, weights are converted to percentages, statistics are calculated, and results are printed. Some of these steps may also be relevant in other programs, and could therefore be prepared as independent and self-contained segments of the program, known as subroutines or subprograms. Subroutines can be regarded as building blocks which can be put together in various ways to produce a range of programs. A subroutine consists of a sequence of instructions or statements each of which specifies a more elementary operation, such as squaring the value of the second data element and adding it to the current total. When compiled, a single statement may be translated into several machine language statements. There is thus a hierarchy of elements in a program. The main program may invoke several subprograms, each subprogram contains many individual statements, and each statement will be compiled into a sequence of machine-language instructions. The levels of the hierarchy are not entirely separate. The main program may consist partly or entirely of statements, and subroutines in assembler language may be required as well as those in a higher-level language.

1.23. Outline of the FORTRAN language

In order to describe high-level languages further, it is convenient to select one to illustrate the main features. FORTRAN is at present the programming language most widely used by geologists. Many excellent books, such as McCracken (1965), are available for those wishing to learn FORTRAN, and courses of instruction, lasting usually from two to five days, are widely available at universities and from computer manufacturers. A programming language cannot be mastered, however, without writing programs and running them on a computer. FORTRAN has some general aspects which it shares with other scientific languages, and a brief survey may be worthwhile, both for those who will later learn a computer language, and for those who will rely on the services of a programmer.

1.24. Words, locations, addresses; names of variables and arrays

It was mentioned above (§1.16) that the numbers and measurements involved in a computation are stored as words in locations in memory. Locations can be thought of as pigeon-holes in which numbers or instructions

D

are filed for reference later in the program. The locations are numbered sequentially, so that each has a reference number, known at its address. An address indicates the position of a location in memory, and enables machine language instructions to store or retrieve one word at that location. In a FORTRAN program, the programmer refers to a location by name rather than by address. When the name is first used, the compiler enters it in a directory, together with the address of a suitable free location, and subsequent mention of that name in the program is taken to refer to that location. Thus, the programmer might decide that a pigeon-hole was required to accumulate the sum of weights of sediment retained in individual sieves in a mechanical analysis. He might choose the name TOTAL to refer to the pigeon-hole, and the compiler would ensure that any reference to TOTAL in the program would indicate the same location. In general, an entity named in this way is know as a variable. In FORTRAN, there are rules about names of variables. They must begin with a letter, consist only of upper-case letters and numbers, and be not more than six characters long. Thus, TOTAL and SIEVE 1 are legitimate names, 1TOTAL, A+B, and PROPORTION are not.

It is not always convenient to give completely separate names to individual variables. For instance, in describing a procedure for summing the weights of sediment in several sieves, each weight must be referred to in the program, but scarcely deserves a separate name. Indeed, a string of different names would be difficult to remember. Array names are therefore used. A name, following the same rules as before, is given, not to a single variable, but to a set of variables, known as an array. Individual elements in the array are identified by a subscript. Thus, the set of weights from all the sieves could be stored in an array called WEIGHT. The weight in the top sieve would then be WEIGHT(1), the next WEIGHT(2), and so on, the subscript being written in brackets after the name. In a similar way, a table of data, with rows and columns, can be stored in a two-dimensional array, with two subscripts, WEIGHT(3, 2), of which the first gives the row number and the second the column number.

1.25. Symbols and expressions; statements; loops and branches

Mathematical calculations are written in FORTRAN as expressions which consist of names, numbers, known as constants, and special symbols representing mathematical operations. The plus and minus signs are used as in arithmetic to denote addition and subtraction. The asterisk * takes the place of the usual multiplication sign to avoid confusion with the letter x. An oblique stroke denotes division and two consecutive asterisks exponentiation. Thus, $(5.0*(A+B*C)**3)/2.0$ reads: five times the value of: A, plus B times C, all cubed, and the result divided by two. Brackets are used as in algebra to indicate which operations are performed first. In the absence of brackets, the order of precedence is exponentiation, multiplication and division, and addition and subtraction. If the brackets were omitted from the above example, therefore, the value of C alone would be cubed and A alone would be multiplied by five.

Expressions represent a sequence of mathematical operations performed on variables and constants. Statements indicate what is to be done with the results. An example of a statement is $X = (Y+Z)/2.0$. This indicates that the values stored in Y and Z are to be inserted in the expression on the right, the result calculated, and stored in the variable named X. In more detail, the number stored in the location named Y is to be added to the number stored in the location named Z, and the result divided by two and stored in the location named X.

The greater part of most programs consists of a sequence of statements that are obeyed in turn by the computer. Normally, each statement is punched between columns 7 and 72 of a punched card. Programs punched on cards are easy to prepare, correct and amend. Comments to explain the arrangement of the program to other programmers can be included at any point, with a C punched in column 1. These are ignored by the compiler. New segments of program can be inserted at any point, and cards with errors can be removed and replaced with the corrected version. The card deck on which the statements are punched is to some extent a physical representation of the sequence of operations which the compiler is to interpret. The logical flow of steps in a calculation is usually more complex than the physical sequence. For instance, certain sequences of steps in a calculation may be repeated several times. Thus, in calculating the average value of a set of measurements, the value of each measurement is added in turn to the running total, and the number of measurements is increased by one. The steps are thus:

```
Set the value of RUNNING TOTAL TO 0.0

Set the value of NUMBER OF MEASUREMENTS to
0.0

Obtain the value of the first measurement

Add the value of the measurement to the
current value of the running total

Add one to the number of measurements
considered.
```

The next measurement is then considered, and the last two steps repeated. It would be tedious and is unnecessary to write the same two statements several times. The part of the program which is repeated is known as a loop.

The statement immediately preceding the loop instructs the computer to DO the following operations repeatedly. It also indicates where the loop ends by referring to the last statement in it. The final statement is identified by a 'label', that is a number punched between columns 1 and 5 on the card, which is not the same number as any other label in the program and can thus be used to uniquely identify the statement. Any statement can be labelled and any statement referred to elsewhere in the program must be labelled. The 'DO' statement which begins a 'DO-LOOP' thus indicates the extent of the loop and the number of times it is to be repeated.

If the average of ten measurements was required, the loop would be repeated ten times. In FORTRAN, this part of the program might look like this:

```
RUNTOT = 0.0
NUMBER = 0
DO 12 I = 1, 10
READ DATA(I)
RUNTOT = RUNTOT+DATA(I)
12 NUMBER = NUMBER+1
```

The value of I is set to 1 the first time that the loop is executed and increased by 1 each time until it is equal to 10. The loop is then executed for the last time, and the instruction following the loop is then obeyed. The final statement in the example could thus equally well have been

```
12 NUMBER = I
```

Another requirement in a computer program may be to follow different paths depending on the current value of a variable. This is known as a 'conditional branch' or in FORTRAN as an 'IF statement'. IF a variable has a negative value, control passes to one statement, IF it is zero to a second, IF positive to a third. An example might be:

```
N = NUMBER-10
IF (N) 13, 11, 12
```

This would result in control passing to the statement labelled 13 if N were negative, and hence NUMBER less than ten, to statement 11 if NUMBER was exactly ten, and to statement 12 if NUMBER was greater than ten. An IF statement could thus be used instead of a DO loop to repeat a sequence of instructions. In the above example, statements 11 and 13 could be at the beginning of the loop and statement 12 could be the statement following the loop. The conditional branch gives programming languages much of their power. It enables the programmer to prepare programs for a problem which is not completely specified in advance. At a number of points in the program, a value is calculated and tested to determine which path the program follows subsequently.

The need for a conditional branch could arise in several ways. In the example mentioned earlier, a program was written to calculate statistics from data obtained by mechanical analysis of sediments. In reading the data, it might be necessary to check whether the end of the data for one item had been reached. A second check could test whether each value was in a permissible range, say between 0 and 100 per cent. Later checks could determine whether sediment had been retained in a sufficiently large number of sieves for the statistics to be valid. A final test might be made to determine whether another set of data remained to be read, or if the end of the data had been reached. At each checkpoint, subsequent processing depends on the result of the test. On the one hand, the program might be terminated; on the other, additional data might be read in. At certain points, depending on the outcome of a test, normal processing might continue, or a message might be written to call the geologist's attention to some undesirable feature of his data.

The conditional branch may transfer control to any point in the sequence of commands or statements which constitute a program. At the end of a branch, it may be necessary to return control to a particular point. For this purpose, an unconditional branch is required, known in FORTRAN as the GO TO statement.

This has the form

    GO TO label

for example

    GO TO 123

which would transfer control from that point in the program to the statement labelled 123.

1.26. Routines; modular programming

The loop and the branch satisfy many of the requirements of controlling the logical flow of a simple program, that is, determining the sequence in which statements are obeyed. With longer and more complex programs, an additional facility is necessary. It is the ability to break up the complete program into a set of separate modules, known in FORTRAN as subprograms or subroutines. A subprogram is simply a set of statements to carry out a specific part of a complete calculation. It may be required once, or several times, in the course of a program. For instance, a subroutine to calculate statistics from one sample of sediment might be required many times during the analysis of a sequence of samples.

A subprogram is presented to the computer and compiled in the same way as a program. It has a name, the naming following the same rules as those for a variable. The name can be up to six characters in length, either letters or numbers or both, with the first character a letter. Reference to the subprogram by name in the main program is equivalent to including the statements which constitute the subprogram at that point in the program. The subprogram is separate and self-contained as far as the computation is concerned, but it is generally necessary to transfer values from the main program to the subprogram, and to pass the results calculated by the subprogram back to the main program. In FORTRAN, there are two distinct types of subprogram, each with a different method of transferring information.

One type of subprogram is known as a function routine. Its purpose is to calculate a single value from one or more values presented to it. It is used in a FORTRAN expression by giving the name of the routine followed by the values it requires in brackets. Thus, SIN(3.1416) would calculate the sine of 3.1416 radians. The statement

$$A = B*SIN(C)$$

would calculate the value of the sine of the number stored in the location named C, multiply the result by the number stored in B and store the result in A. A number of function routines, including SIN, are available with the FORTRAN compiler. The user need not write his own programs for these, but can use the ones already available.

The other type of subprogram is known as a subroutine. Any number of values may be transferred to it and returned to the main program from it. It can communicate, that is, transfer information, with the main program in one of two ways. A set of variables or arrays can be specified in brackets after the subroutine name, in what is known as an argument list. Alternatively, a statement called a COMMON statement can be included in both the main program and the subroutine, giving the names of variables containing values that are to be transferred between them. The subroutine is invoked in the main program by a CALL statement which has the form

$$CALL \; name \; (argument \; list)$$

For example, if a subroutine named HIST had been prepared to plot a histogram, the statement

$$CALL \; HIST \; (DIST, \; N)$$

might call the subroutine HIST to plot a histogram of a frequency distribution stored in the first N elements of the array called DIST. The effect of a CALL statement is similar to that of inserting the statements of the subroutine at that point in the calling program. The principal difference is that names of variables other than those in a COMMON statement have a different meaning in the subroutine and the calling program, and communication must therefore be through COMMON variables or an argument list. The subroutine need not be called from the main program, but can be called from another subroutine.

There are various reasons for using subroutines. A long program of more than say a hundred statements may not compile as efficiently as a smaller one. It may be preferable to break the program down into a set of modules, since each subprogram is compiled separately. It is also easier to locate errors in a small module which can be tested on its own, than in a long program. Certain operations may be difficult or impossible to code in FORTRAN, since it is a language designed for programming scientific calculations. Such operations can be programmed as subroutines in a lower level language, such as ASSEMBLER, and called in a FORTRAN program. In an involved programming task, the use of subroutines makes it possible to schedule the work effectively, by assigning the preparation of separate groups of subprograms to different programmers. Perhaps the most important reason for using subprograms, however, is that the same set of mathematical operations may be required in various contexts. They can be programmed efficiently once as subprograms, and called whenever required. Most of the applications for which the geologist uses the computer overlap to some extent with earlier programming work. It is important, therefore, to be able to utilize existing programs, and adapt them for varying requirements. Their structure, in terms of routines and subprograms is thus highly significant.

1.27. Input and output statements; format statements

Programming languages must be capable of specifying input and output of information from the computer, such as controlling the reading of data cards and printing results. Input and output are handled in FORTRAN by READ and WRITE statements. The former controls the transfer of information from the form in which it was prepared or stored to the form in which it is used within the computer. An example would be reading characters from punched cards and storing them, assembled as words, in main memory. The WRITE statement transforms results produced by the program to a printed version that can be read by the user, or to back-up storage such as disk or magnetic tape, where the results will be available later. Three kinds of information are relevant in an input or output instruction. One is the input or output device to or from which information is to be transferred. Another is the type and arrangement of the information, known in FORTRAN as its format. A third is the variable names by which the items of information are known in the program.

In READ or WRITE statements, the input or output device is specified by a number. The card reader is usually referred to as channel 5, and the lineprinter as channel 6, but the numbers used depend on the particular configuration. It is convenient therefore, to refer to these devices by variable names which are used throughout a program, and to set the variables to appropriate values at the beginning of the program. Only these initial statements need then be altered to run the program on another computer. The format is described in a separate statement, described below. A format statement is always labelled, and referred to by the appropriate label number in the READ or WRITE statement. The device number and format appear near the beginning of the input or output statement, following the word READ or WRITE in brackets, and the rest of the statement consists of a list of the variables in which the information is stored by the program. The following are examples:

READ (5,100) A, B, C, D

Transfer to the variables named A, B, C and D the first four items available from device 5, according to the format listed in statement 100.

WRITE (IOUT, 2) Y, EM, C

The values stored in the variables Y, EM and C are to be written in the format specified in the statement labelled 2, on the device with number stored in the variable IOUT.

The fact that the format is separate from the READ and WRITE statements has the advantage that formats need not be repeated if they are used by more than one READ or WRITE statement. The format statement describes the type and arrangement of a sequence of fields. Each field corresponds to one item in the variable list of an input or output statement, and is thus generally the smallest item of information that can be usefully processed as a single unit. A field may contain an integer number, a real number, consisting of a whole number and a decimal fraction, or a sequence of characters. These types of fields are denoted I, F and A, for integer, floating-point and alphanumeric fields respectively. The field width, that is the number of characters occupied by the field, is also indicated, and the position of the decimal point in a floating-point field. The format statement is required most often for reading data from punched cards or listing results on a lineprinter. In both devices, it is convenient to ensure that the data are automatically tabulated by assigning each field a fixed length and fixed position. The data are then said to be in fixed format, normally with each field corresponding to exactly one word in core store. Examples of format statements follow:

```
100 FORMAT (F10.3, F8.6, F4.0, F4.0)
```

If referred to by the READ statement given in an earlier example, this format would indicate that four floating-point numbers are expected, the first occupying the first ten columns of a punched card, the second the next eight columns, the third the next four, and the fourth the next four columns. In all, therefore, the first 26 columns of the card would be read. The field widths are indicated by the number before the decimal points in the format statements. The numbers following the decimal point, such as the 3 in F10·3, indicate where the decimal point falls in the field itself. In the first field of the punched card the last three columns are assumed to follow the decimal point, in the second field, it falls before the last six columns of the field, and in the third and fourth fields, it lies at the end of the field. No decimal points need be punched on the data cards themselves, but if a point is punched, it overrides the implied position given by the format statement.

Where adjacent fields have the same format, it need not be written out each time. A shorter version of the above statement is:

```
100 FORMAT (F10.3, F8.6, 2F4.0)
```

The number before the F, I or A, indicates the number of times that the field is repeated, one being assumed if no number is present. The letter X is used to indicate blank columns on cards or lineprinter, on input or output, and is preceded by a number indicating the number of adjacent blank columns. Text information for labelling output can also be introduced in a format statement. The text is enclosed in single quote marks, or is introduced by the letter H preceded by a count of the number of columns of text. An example is:

```
2 FORMAT (34H PARAMETERS FOR THE FIRST
    EQUATION, 2X, F5.2, 4HY = , F5.2, 4HX
    + , F5.2)
```

Format statement 2, in conjunction with the WRITE statement

```
WRITE (IOUT, 2) Y, EM, C
```

would result in the following line being printed, assuming that IOUT is set to the device number of the line-printer, and that Y, EM and C have the values $27 \cdot 4$, $13 \cdot 26$ and $0 \cdot 02$ when control passes to the WRITE statement.

```
PARAMETERS FOR THE FIRST EQUATION
27.40Y = 13.26X + 0.02
```

The following conventions of FORTRAN cause confusion to some users. A blank field read in numeric format is stored as zero. It is not possible, therefore, to distinguish by a blank and zero between a missing value and an actual measurement of $0 \cdot 0$. Instead, a missing value code (see $2 \cdot 14$) can be used. On printing, blanks rather than zeros precede a value to make up the full field width. Thus, the value $2 \cdot 14$ is printed in F6·2 format as $2 \cdot 14$, not $002 \cdot 14$, whether or not there were leading zeros in the original data. Printed values are truncated rather than rounded, so that if the stored value were $2 \cdot 14999$, the number printed in F6·2 would nevertheless be $2 \cdot 14$.

As FORTRAN was designed for scientific and mathematical programming, data are generally presented and results returned in fixed format. Many geological applications, however, involve handling text information as well as numbers, and fixed format is not always suitable. Some FORTRAN compilers can accept free format data, in which fields are separated by special characters, such as blanks, rather than always being punched in the same position on each card. Compilers with this facility are naturally more complex, and may therefore take longer and use more core store to compile a program, whether or not free format data are involved. An additional drawback is that if special features of compilers, such as this, are used, transfer of programs from one computer to another becomes more difficult. If programs are prepared for general use, or for use over a long period of time, which may mean transfer from one computer to its successor, then it is advisable to use only the minimum set of FORTRAN facilities specified as an international standard by the American National Standards Institute. A summary account of ANSI FORTRAN and of the differences between compilers is given by the National Computing Centre in 'Standard Fortran' (1970).

Only a brief outline of some features of FORTRAN has been given above, and this, of course, is quite inadequate for the geologist who wishes to write his own programs. However, FORTRAN is not a difficult language to learn, and many excellent courses and textbooks are available. The only way to master a language is to use it, and it is advisable to time any detailed study of FORTRAN to immediately precede an intensive period of programming. One purpose of the above account was to introduce the geologist to a specific language and call his attention to certain features of FORTRAN which are likely to be important to him whether or not they are mentioned in an elementary course. Another purpose was to describe features which most scientific programming languages have in common. Clearly, it must be possible to control input of data and output of results. The loop, jump and conditional transfer are features of most scientific programming languages. Variables, arrays, expressions and statements are generally necessary in some form, and the ability to segment a program in a hierarchical manner, achieved in FORTRAN by subroutines, adds greatly to the power of a language.

1.28. Flowcharts

Some understanding of these aspects of a language is needed even by the geologist who has his programs written for him. A simpler language than FORTRAN is, however, available, in which it is possible for the geologist to express his strategy for solving a problem. It is not a computer language and cannot be compiled directly. But it can be used as a means of communicating with a programmer, or for setting out a preliminary analysis of a problem before developing the program in detail. Loops, jumps, conditional branches and subroutines can be shown, and the sequence of flow of control can be clearly expressed. A program prepared in this language is known as a flowchart.

Flowcharts are generally produced at an early stage in the preparation of a program, in order to describe the logical relationships and sequence of steps. The flowchart is relevant whatever computer language is finally used. Perhaps it is most useful where a systems analyst or geologist is responsible for specifying the program, which is subsequently encoded by a programmer. If the geologist prepares his own program, he may find that a flowchart is unnecessary. But if the program will be used later by others who must understand its structure in detail, a flowchart may be helpful. Where one is not provided, the user may have to draw up his own in order to understand the structure of the program.

Flowcharts can represent any level of detail from a broad outline of the required procedures, to detailed instructions corresponding to statements in the computer language. The steps in the program, at the chosen level of detail, are written within boxes the shape of which conventionally represents the type of step, as shown in Fig. 11. The boxes are joined by arrows to indicate the sequence in which the steps are followed. Loops can thus be shown graphically by arrows leading back to an earlier step in the sequence. Points of decision are of particular importance, and a conditional branch is shown by a diamond-shaped box with alternative paths from it, selected according to whether or not the condition within the box is satisfied. Sequences of operations which lie outside the program of immediate interest, such as subroutines, can be indicated in an appropriate box, and cross-referenced by name to a separate, more detailed flowchart.

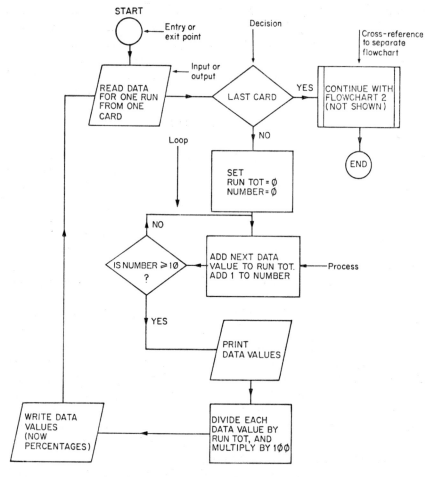

Fig. 11. Flowchart illustrating some steps in a simple program.

1.29. Operating systems and job control language

The software on a computer has to be capable of doing more than just compiling and running a program. In order to handle satisfactorily a set of programs for several users, the computer has to accept jobs from many sources, including remote terminals, to ensure that programs supplied by the manufacturer, such as the assembler or the FORTRAN compiler, and routines, such as those for calculating sines or square roots, are available to the programs that require them. It may have to link together segments of program, such as a utility routine for sorting a data set and the main program written by the user, and to link these in turn with datasets retrieved from tape or disk. It has to ensure that the correct tapes and disks are mounted and available to the system, that the jobs are scheduled to make efficient use of the system resources, and that each job is timed and a statement of charges prepared for the user. It must also check the running of the job against limits of maximum running time and lines of output, as it is entirely possible for a program error to cause the same page of output to be printed repeatedly or a calculation to repeat continuously in a loop that has no exit. In addition, the results of the computation must be sent to the appropriate terminal or output device, and be clearly identified in order that they can be returned to the correct user.

These tasks are handled by the operating system of the computer. On a large, general-purpose computer, the operating system is inevitably complex, and a significant cost overhead. On a minicomputer, a much smaller and simpler operating system is used, and if the user or operator is on hand when the job is run, precautions against over-running may be unnecessary. The information which the operating system requires for each job is supplied by the user in job control language (JCL), the form of which varies from one computer manufacturer to another. The job control statements for a job which is run on a large computer can be quite complex, and their preparation is time-consuming. However, for a simple, routine job, the task may be simplified by the 'default options' supplied by the operating system. The default option means that if the user does not explicitly state a value for a control parameter, the operating system supplies a standard value. Another aid is the availability in many systems of sequences of job control statements which have already been prepared, and can be invoked by name, rather like invoking a sequence of statements by calling a subroutine in a FORTRAN program. However, if the job is not of standard type, perhaps requiring access to a specific disk or tape file, then the appropriate job control language statements are required.

1.30. Program libraries, packages and systems

Most of the widely used computer procedures have been prepared in many versions by many programmers. It is possible to exchange information on methods of problem-solving by exchange of programs, a topic considered later in the chapter on communications. This shared pool of information can take several forms, such as program libraries, packages and systems. A library of programs or subroutines, such as the utility programs supplied by the computer manufacturer for such purposes as copying tapes or editing files, and routines in high-level languages like FORTRAN, are available for the user to incorporate in his own program, or to amend for his own purposes.

Access to a comprehensive library of tested routines gives the programmer considerable flexibility in constructing the programs he wants. However, an appreciable programming effort may be required to ensure compatibility between the segments of the program. Packages are available within limited subject areas, such as statistics or graphic display, where the programs and subroutines are carefully designed to link together, as modules or building bricks from which the user can construct the sequence of operations that he requires, within the framework of a main program and job control language supplied or amended by the user. This approach takes full advantage of the modular structure of a programming language, allowing the user to express his ideas in larger modules than a single program statement, gaining considerably in convenience at the expense of some flexibility.

A further step in user convenience is an integrated system which has been designed for a limited range of tasks, and may operate under the control of an executive system (see Jones *et al.*, 1976). The commands may be simple mnemonics or English words, or the system may operate on the menu technique where a list of alternatives is presented to the user, from which he selects one by typing its number or code. The executive program responds by supplying the necessary programs and job control language from a file on the computer, and building and running the task without further user intervention. Although it would be possible to allow a programmer to make changes at any level, it is usual in a system of this kind to give the user access to the high-level commands only, to prevent unauthorized modifications and rapid deterioration in the quality and consistency of the system.

In summary, software of varying degree of complexity is available to the user to enable him to specify the activity he wishes the computer to perform. At the simplest level, a command may control the movement of a few binary digits in store. At a complex level, it may invoke programs, data and job control statements to perform a complicated task, like drawing a contour map for a set of points or performing a multivariate analysis of a set of data. With increasing complexity of software, the user relies increasingly on methodologies, techniques, procedures and implementations designed and developed by others. Although this has the limitation that he must accept decisions made by other workers in different contexts, it has the major advantage that, with little effort on his part, he can bring to bear on his problem the results of many man-years of highly skilled and specialized work, represented in the high-level languages, programs and systems by which the user can control and utilize the power of the computer.

2. OBSERVATION AND REPRESENTATION

2.1. The objectives of data collection and recording;
data as illustration and as evidence

In considering the various subsystems of geological investigations, the chosen starting point was observation of the real world, the eventual source of all geological knowledge. The raw data collected by the observations of geologists as a whole are clearly too numerous for any single individual to comprehend them. The data are therefore recorded and organized, and the amount of information reduced step by step as scientists attempt to find common ground in shared knowledge, not of individual observations, but of broad underlying patterns, expressed as hypotheses, theories and general concepts and principles. At an early stage in this process, computer methods can come into play. The first subsystem, as considered in this chapter, includes observation, the collection and recording of data, and their representation and organization within the computer. At the risk of stating the obvious, some general features are considered which may be taken for granted in a conventional study, but can assume some significance where new methods are being introduced.

The geologist's objectives in observing and recording data are varied. Data collection is an active occupation, in which progress, in the form of completed maps and field notebooks, can be readily measured. There is a danger that such activity becomes an end in itself, without consideration of reasons and methods. But the work is usually undertaken with clear objectives in mind. In a research project, field work may have the aim of testing an existing hypothesis. Economic investigations may aim to discover the presence or distribution of specific minerals. Geological investigations can have many detailed objectives, which may overlap. The collection of data for each objective separately would create much redundant information. It may therefore be more efficient for many purposes to carry out systematic general geological surveys, and in many countries an organization exists for this purpose. Typically, the data from a general survey are made available after some condensation and analysis, but still in considerable detail. A geological interpretation may accompany such data, but usually in general terms which do not prevent testing a range of hypotheses against the data.

Recording of data is concerned with representing what is observed and communicating it to others. In geology, a full account of an observation may be impossible. A geologist might immediately recognize a hand specimen as coming from a particular formation, but be quite unable to do so from the most detailed written description. The geologist must be highly selective in what he observes, and more so in what he records. A lifetime is too short to describe even a hand specimen in full detail. A geologist may use his own records to remind himself of details of past observations (Where did I see that red and green striped mudstone?) without it being possible for others who did not share his observations to use them in this way. Data may even be recorded to help in making systematic observations, with no intention of referring back to the data. It is thus possible that only part of the recorded information can be communicated to a wider group of geologists who do not share the same background or experience. The geological interpretation of an area may depend more on what was observed than on what was recorded. In these circumstances the geologists who made the observations may be the only ones able to interpret the data correctly, and any computer analysis is for their benefit alone. The conclusions, rather than the data, are communicated more widely, and data may be quoted as illustrations of a theory, rather than as evidence from which others may draw their own conclusions.

2.2. The effect of computer methods on data collection

Computer methods have wider implications where a formal mathematical notation can be used for describing observations, and data can be collected according to a reproducible sampling scheme. They make it possible for one scientist to re-examine in full detail data collected by another. The level at which communication can take place is thus moved from summary and interpretation to direct records of observations, and the integration of the detailed observations of many workers becomes possible. The ability of the computer to analyze large amounts of information in detail means that the quantities of data obtained from, say, data loggers or automatic methods of geochemical analysis, can be fully examined. These benefits are more readily achieved in geophysics and geochemistry than in the more traditional areas of field investigation. However, existing descriptive languages, such as English, have their drawbacks, and experience with computer methods may already be leading to improved notational systems for geological description. Their potential value can be judged by attempting to describe six bars of a symphony or a knitting pattern in standard English without the benefit of a specialized notation. Computer techniques for the collection and description of data and their organization and management depend on a number of mathematical concepts, some of which will now be described.

2.3. Set theory; sets, subsets, union and intersection

Data are a collection of related items of factual information, generally derived from observation. In an abstract, mathematical sense, it is possible to study the properties of a set of items and the form of the relationships between them, independently of their substantive significance, that is, their content and meaning in terms of the real world. The abstract mathematical approach involves parts of set theory, formal logic and topology. The mathematical background is described in various textbooks, such as Stoll (1961). Many of the conclusions, however, are intuitively obvious, and for present purposes, an informal explanation of a few terms and a statement of some relevant principles should suffice.

A set is a collection of entities or objects, each of which is known as a member of the set. A subset is defined as part of a set. Thus, a collection of thickness measurements of beds of sandstone encountered in a bore hole might comprise a subset of the set of thickness measurements of all beds penetrated by the borehole. Individual thickness measurements are members of the set and, in some cases, of the subset. As with systems and subsystems, the question of what constitutes a set or subset is decided on the basis of what is appropriate and convenient. Thus, if a geologist were studying a collection of fossils, he might define his set as, say, the fossils in cabinets 1 to 5 in Room G6. If he were studying variation in properties between classes, he might then define brachiopods as one subset, trilobites as a second, and so on. Subsets are not mutually exclusive. The subset 'marine fossils' and the subset 'brachiopods' might contain the same members. The subset 'fossils more than one inch in length' might include some, but not all, of the members of the subset 'marine fossils'. If the geologist were to investigate the stratigraphic relationships of the fossils, he might introduce a new set, namely, the stratigraphic formations from which the fossils were collected. A fossil might thus be related to the member 'Tensing' of the set 'formations'.

Two sets, or subsets, are said to intersect when they overlap so that certain items are members of both sets. This situation can be represented diagrammatically as in Fig. 12. The area of overlap is known as the intersection of the sets. Thus, large brachiopods are members of the subset formed by the intersection of the set 'large fossils' and the set 'brachiopods'. Each member of the intersection of two sets is a member of both sets. The area which contains all of both subsets, on the other hand, is known as the union of the subsets. Thus the union of the subset 'large fossils' and the subset 'brachio-

E

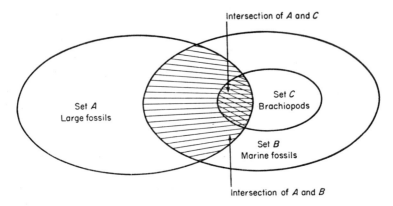

Fig. 12. Venn diagram illustrating the union and intersection of sets. The set of marine fossils contains brachiopods as a subset.

pods' would contain members which were either large or brachiopods, and could be both.

None of these ideas from set theory is immediately helpful to the geologist studying fossils. They do, however, provide a general framework in which to describe data retrieval. It would clearly be wasteful to have one suite of programs for fossil data, another for rock specimen data, a third for geochemical data and a fourth for literature retrieval if all the programs involved precisely the same operations in the computer. It is desirable, therefore, to be able to express the operations in general terms, dealing with sets and subsets, unions and intersections. If the geologist can consider his own substantive problem and then define his data processing requirement within a general framework, the task of finding and using appropriate programs is simplified.

2.4. Properties and relationships; formal logic

A set of data elements may define the properties of a set of objects. For example, data might be recorded for the fossils mentioned above, indicating their identification number, class, genus, species, environment, size, locality and formation. It would be possible to select marine fossils in the set by determining which members satisfied the relationship: environment is marine. At a less simple level, all marine bivalves could be retrieved by determining which members satisfy the relationships: environment is marine and class is bivalve. More complex relationships might be: fossils which are either plants, brachiopods or bivalves and from a marine environment but which do not come from the Ercu Mountains and are very small. This type of selection is not unusual in working with large data files or performing a detailed computer analysis of causes of variation of some property, say, shell thickness. It is difficult to express the requirement in simple, unambiguous English. For example, does the sentence above indicate that a large brachiopod from the Ercu Mountains would qualify?

The language which makes it possible to express such relationships precisely and unambiguously, regardless of the data content, is that of formal logic. The four elements of the language that are immediately relevant are: 'and', 'or', 'not', and '()'. Brackets are used to indicate the scope of relationships. Thus, not (large and marine) signifies that the property (large and marine) does not apply, while 'and' indicates that both the properties must occur together. On the other hand, 'or' indicates that at least one property applies. Thus, (large or marine) would signify that a large fossil, a marine fossil, or a large marine fossil would be relevant. The 'exclusive or', sometimes written 'xor', implies one or other but not both. In complicated selections, phrases within brackets may themselves be placed within brackets. Without departing far from normal English, these conventions can be applied to the example: Fossils which are ((plants or brachiopods or bivalves) and are marine) but not (from the Ercu Mountains and are very small). If the brackets were placed differently, the meaning would naturally alter; for example: (plants or brachiopods) or (bivalves and marine).

Perhaps the main application in geology of ideas from formal logic is in the precise statement of criteria defining subsets of information. In mathematical applications, greater generality is achieved by the use of symbols for relationships and logical operators. It thus becomes possible to examine the form of the logic without regard to the meaning of the words. The appearance of phrases like $s \in T \cup V$, or $(\sim S \vee T) \vee (\sim T \wedge V)$ is unfamiliar to most geologists, although they are not difficult to understand. The first phrase signifies that s is a member or an element of the set formed by the union of the sets T and V. The second phrase means ((not S) or T) or ((not T) and V). Again this may seem meaningless, but if in a particular application S means small, T means marine and V means bivalve, the phrase would indicate (either (not small) or marine) or a non-marine bivalve, and its significance becomes clear. The advantage of the symbolic representation is that the emphasis is placed on the relationships, and obscurities, contradictions and possible simplifications may be revealed. More familiar symbols are used in most computer languages, and are described in the appropriate manuals.

Relationships are also important in many geological hypotheses, and can be expressed in formal terms. The hypothesis that in a particular formation, coarse sandstone beds with rootlets do not contain fragments of marine limestone, for example, could be expressed and tested against an appropriate data file.

2.5. Data structures; arrays, matrices, trees and networks

Within data sets, relationships exist which are inherent in the data themselves. The relationships between data items are an essential part of the data, and are generally established when the data are recorded. The pattern of relationships within a set of data items is known as the data structure. The structure of data has an important bearing on the ways in which they can be processed and analyzed.

A single data item, the value 2·45, for example, has no significance in itself. Even the fact that 2·45 is a thickness in meters is meaningless. A sequence of thicknesses of successive beds in a specified borehole, however, could have geological significance. The individual values are no longer unrelated items of information, but are tied together by their position in sequence. The links between the items, implied by their sequential position, constitute the structure. This particular type of data structure is known as a 'string'. Diagrammatically, it can be depicted, as in Fig. 13, by a single line joining the items. In the example of a borehole, other information would generally be attached to each thickness. For instance, the lithology, maximum grain size and color might be recorded for each bed. An individual data item, such as 'sandstone', would then be related in two directions. It is first linked to the bed to which it refers, and secondly to other records of lithology. Information of this kind might be recorded as a table in which each row refers to a bed, and each column to a property, as shown in Fig. 2. Diagrammatically, this structure, known as an 'array', can be depicted by a grid of lines (Fig. 13). The rows are sometimes known as 'records', the columns as 'fields' or 'variables'. Arrays with rows and columns are known as two-dimensional arrays. Strings, by analogy, are one-dimensional arrays or vectors. Three-dimensional arrays are less often encountered in geology. An example might be data about the chemical composition of water from various aquifers (a two-dimensional array) recorded each day for a period of one month. Data from such a three-dimensional array could be written as a set of tables, all of similar format.

A structure of particular importance in data analysis is known as a matrix. It is a two-dimensional array of quantitative data. The study of matrices is a specialized branch of algebra, namely, matrix algebra. It offers a compact notation for expressing complicated operations, many of which are basic procedures in statistical analysis and may correspond to easily identifiable steps in a computer program. A complete matrix is represented by an upper-case letter in bold-face type, thus: A. Each of the individual quantitative values in the array is known as an element of the matrix. An element is denoted by a letter in lower-case (the same letter as the name of the matrix), with two subscripts indicating the row and column of the matrix occupied by

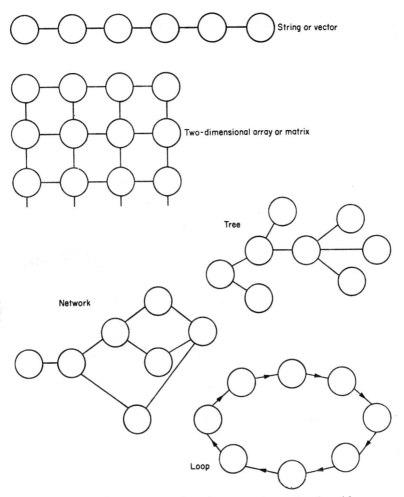

Fig. 13. Some data structures: the string, array, tree, network and loop.

that element. Thus $a_{4,\,2}$ is the element in the fourth row and second column of the matrix A. The notation corresponds directly to the FORTRAN representation A (4, 2). If the matrix is written out in full, with the value of each element given, it is usual to enclose the matrix in square brackets. Thus

$$A = \begin{bmatrix} 12\cdot4 & 3\cdot2 & 10\cdot7 \\ 14\cdot6 & 2\cdot8 & 9\cdot3 \\ 15\cdot8 & 3\cdot9 & 8\cdot8 \\ 17\cdot4 & 4\cdot2 & 6\cdot9 \end{bmatrix}.$$

The number of rows and columns determines the size of the matrix. The matrix A as written above is a 4×3 matrix, read as four by three, that is it contains 4 rows and 3 columns. The notation thus offers a compact and unambiguous notation for referring to entire sets of numbers, and also for describing operations and calculations involving entire matrices.

The details of matrix algebra are not, perhaps, immediately relevant to data collection. However, it is an important point that many techniques of data analysis can be expressed in matrix algebra. If the data can be written in matrix form, they will probably be amenable to subsequent mathematical analysis. Intuitively, if each row in a matrix corresponds to one item, and each column to one property, then a complete matrix has the advantage that measurements on all items correspond exactly, and thus relationships among all the properties can be studied.

It is not always appropriate to record the same types of information in each record of a file. For example, one bed penetrated by a borehole might contain remains of plant fragments, while the underlying bed contained grains of purplish quartz. A lower bed might consist of conglomerate with convoluted fragments of light grey, thinly bedded siltstone. It would not be practicable to allow for all these eventualities in a description by leaving space for them in a table. If the links between successive items of information are depicted diagrammatically, they form a branching structure, known as a tree (see Fig. 13). Thus, the items of information 'light grey' and 'thinly bedded' might both be linked to 'siltstone', while siltstone might be related to 'fragments' and this in turn to 'conglomerate'. A sequence of tree structures, such as descriptions of a sequence of beds, is sometimes known, for obvious reasons, as a forest. Information originally recorded as a tree can be transformed into an array, simply by leaving a position in the array available for every possible node in the tree. It is then likely that only a small proportion of array positions would contain information. This is known as a sparse array. The operation of transforming data from one structure to another is known as mapping. It may be performed in the computer in order, for example, to alter data from an arrangement convenient for recording to one convenient for storage and retrieval.

There are other data structures which are less frequently encountered in geology. The network, for example, differs from the tree in that instead of an expanding, branching structure, each data item, or node, can be linked directly to any other in the network. The links may or may not be directional. An example of a directional network is shown in Fig. 14. The data items are events in the geological history of an area. The links show the sequence of events in time. Thus an arrow from event A to event B indicates that A happened before B. The flowchart diagram, mentioned in the previous chapter, is an example of the representation of information as a network, a

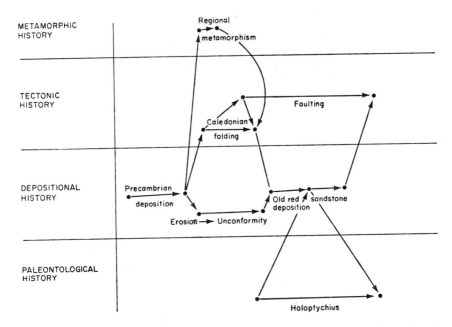

METAMORPHIC
HISTORY

TECTONIC
HISTORY

DEPOSITIONAL
HISTORY

PALEONTOLOGICAL
HISTORY

Fig. 14. A directed network showing events in an area. Reproduced from Loudon (1971).

technique which can be applied to many systems which describe operations or events. A tree can be regarded as a special case of a network, with branching in one direction only. Another less common structure is the ring, which can be regarded as a special kind of string, or as another type of network. It is again directional, but has no branches, and is closed (see Fig. 13). That is, if the arrows are followed from any point, they eventually lead back to the starting point. The locations of points on a closed contour line, for instance, form a ring.

These structures have an important role in data storage and retrieval (see Loudon, 1971). Data structures are general concepts which do not depend on the actual information to which they refer. The data items might concern lithologies, fossils, maps, bibliographical references, chemical analyses, physical properties, orogenies, grains of sand or continents. Examples of the relationships between items might be as follows.

1. Links from the measurement of a property to the object on which the measurement was made. Example: table of geochemical analyses.

2. Links from the object described to the adjacent object in space or time. Example: sequence of beds described from a borehole.

3. Links from the object described to the description of its component parts. Example: description of sand grains within a bed of sandstone.

4. Links from an event to another event known to precede it in time. Example: an account of the historical geology of an area.

5. Links within a geological process between events and other activities which must precede them. Example: rain must fall for water to flow in a stream, and the water must flow before a pebble can be moved.

2.6. Types of data element; quantitative and qualitative

In computer data analysis, the structure of data is an important consideration. Equally important is the nature of data items. If a set of measurements refers to bed thickness, the average thickness can be calculated. If the data items refer to sedimentary structures, it is meaningless to calculate the average of two graded beddings and three ripple drift laminations. Quantitative and qualitative data are different in kind. It is, of course, possible to represent qualitative data by numbers. Graded bedding could be indicated by 1, cross-bedding by 2, and so on. It would then be numeric data, but would still not be quantitative, and the average of several classes would still be meaningless. The distinction between data types is summarized by Krumbein and Graybill (1965). Whereas quantitative measurements assign a data item to a precise position on a continuous scale, qualitative information merely assigns it to a class. Classes may or may not be ranked, that is, arranged in sequence. For example, the classes conglomerate, sandstone, siltstone, and mudstone form a sequence in terms of grain size. The classes quartz, feldspar and magnetite, on the other hand, cannot be so readily arranged in a meaningful order.

Quantitative data themselves, are of various kinds. Thickness measurements, for example, refer to a scale which has a true value of zero. Temperature in degrees Celsius, on the other hand, has an arbitrary zero. It is meaningless to say that 40°C is twice as hot as 20°C. Many geological measurements, such as modal analyses, are percentage data. By definition, they add to a constant total. This, too, has an effect on the methods of analysis that are appropriate. In the extreme example of a two-component system, the proportion of one component, say, of sand grains in a sandstone, precisely determines the proportion of the other, the component which is not sand grains. Plotting the weight of one against the other would reveal nothing more than the existence of an obvious relationship, see Fig. 15. With three or more components adding to a constant total, this relationship is masked, but still exists. An extensive literature has developed in geology on the misconceptions that can arise from careless analysis of 'closed-number'

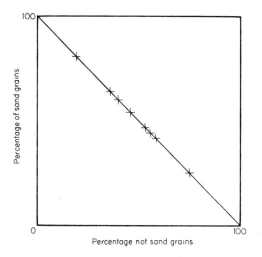

Fig. 15. A two-component closed system.

systems (see Koch and Link, 1970). Orientation data also give rise to problems in analysis. If orientations of currents depositing a shoestring sand at four points were 360°, 40°, 100° and 300°, the average direction is not (360+40+100+300)°/4 or 200°, but is in fact 20° (see Fig. 16). The measurement of 360° could equally well have been recorded as 0°, but (0+40+100+300)°/4 would give another inaccurate average of 110°. These points are considered in the chapter on interpretation.

Since these aspects greatly affect the extent to which data can meet the

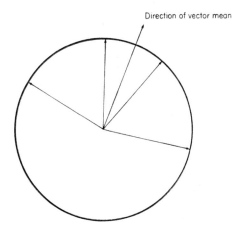

Fig. 16. The average of four orientation measurements, given by the vector mean.

objectives of processing, it is important that they be considered at an early stage in an investigation, before detailed data collection begins. Because certain data structures and types of data are particularly suitable for computer data analysis, they occupy an important position in both the theory and practice of computer applications in geology. In turn, they influence the collection and recording of data.

2.7 Sampling theory; random, multilevel and sequential sampling

During a geological investigation, only a small part of the material about which conclusions are drawn is actually observed, and only a small part of what was observed is recorded as data. For example, a geologist may draw conclusions about a sedimentary basin hundreds of miles across on the basis of a few hundred boreholes each a few inches in diameter. It is quite probable that the original cores and cuttings were not available, and that conclusions were drawn on the basis of written descriptions prepared by several other geologists. The geologist who collects new data with specific objectives in mind is better able to ensure that the entire investigation, including data collection, is carried out as efficiently as possible. But even where existing data are used, some selection is likely to be necessary. The branch of mathematics which is concerned with this problem is sampling theory, an aspect of statistics. Common sense and knowledge of the geological background are always more important than mathematics in a geological investigation, and indeed the first rule of statistical applications is never to lose sight of the subject matter. Nevertheless the validity of conclusions depends on the appropriateness of the data collection technique, particularly where a complex computer analysis is attempted.

An entire set of objects under consideration, such as those in a system which is the subject of a geological investigation, is known as a universe or 'population'. The original use of this word in statistics was presumably to refer to a population of individual human beings. As a technical term, it has gained a wider meaning, and statisticians might refer to a population of males over twenty-five, a population of sand grains, or a population of measurements of sediment movement in the Mississippi River. In geology, the population of interest, the 'target population', is not necessarily the same as the available population. In a study of the Carboniferous rocks of an area, only outcrops might be accessible to observation. Of the remainder of the target population, deeply buried strata might be accessible in theory, if not in practice. In other areas, where the Carboniferous rocks were long since eroded away, there is no prospect of ever making direct observation of their properties.

Even where the entire target population is accessible to observation, it is seldom necessary, desirable, or even possible, to examine it all fully. Instead, a selected part of the population, known as a sample, which is considered in some way to be representative of the population as a whole, is observed. Sampling theory is concerned with techniques for economical and efficient selection of a sample appropriate to the objectives of the investigation. Even if precise objectives are not fully defined before an investigation begins, the reasons for carrying out the study and the relative importance of different aspects will be known in advance. This can perhaps be made clearer by giving a number of examples of possible objectives.

1. A geochemical reconnaissance might be undertaken to attempt to establish possible locations of economic mineral deposits.

2. A borehole might be drilled to look for an oil reservoir in an anticline indicated by seismic survey.

3. Samples might be collected and analyzed to determine whether a significant change in geochemical content occurs near the margin of a granite intrusion, and if so, whether such changes relate to the granite.

4. Borehole logs might be examined throughout an area to determine whether the nature of cyclicity in a sedimentary sequence changed between the shallow and deep water areas of the basin of deposition.

In the first example, the sampling scheme has the objective of ensuring that, if an economic deposit exists, there is a high degree of probability of gaining evidence of its existence and whereabouts from the sample. In the second case, a hypothesis that there is an oil pool in a particular area is to be tested. A test of a more complex hypothesis is required in the third example, which calls for estimation of various statistical parameters of several populations on the basis of a limited sample. The fourth example again involves complex hypothesis testing, but with the major difference that the available sample, namely the borehole records, is determined by factors outside the control of the investigator.

The introduction of computer methods has a bearing on each of these examples. Computer simulation, by allowing a detailed study of possible geological situations, can lead to more effective search techniques, required in examples 1 and 2. The possibility of holding and communicating raw data by means of a computer data bank can lead to the same data being used by a number of geologists with different objectives. The results could be totally misleading unless careful attention is paid to sampling problems, both in collecting and using the data. Computer analysis provides an opportunity for extracting as much information as possible from data, but cannot proceed far unless the sample is appropriate.

In the first and second examples, the aim is to search for a defined situation. Background knowledge may indicate areas where success is most probable. One formation, or one area of tectonic activity, might be known to be more favorable to mineralization than another. Economic considerations might indicate the minimum extent of the deposit or reservoir that is sought and geological information might indicate the shape and orientation that is most likely to occur. Knowledge of the geography of the area would help to decide the pattern on which it would be easiest to collect data. In geochemical reconnaissance, samples might be obtained most cheaply by following rivers or roads. In drilling a well, ease of access to the site may be an important consideration. Such a system may be too complex to allow clear intuitive conclusions about the sampling scheme, but the objectives are clear and many of the variables can be quantified. The costs of exploration may be sufficiently high to justify detailed preliminary investigation of the exploration procedure. The most satisfactory way of investigating the probable cost and outcome of the various possible exploration procedures may be to represent aspects of the system, or 'simulate' it, mathematically.

On the basis of assumptions about the relative probability of occurrence of various geological situations, and a range of exploration procedures that could be followed, it may be possible to determine which procedures are most likely to be successful. With a single, precise hypothesis it may be possible to draw conclusions from a graphical simulation, as in Fig. 17. The real world is usually more complicated, and a large range of types of hypotheses must be considered, rather than a single instance. To examine these comprehensively by manual methods could be impossibly time consuming, and computer simulation could be an attractive alternative. Computer simulation is considered in detail in the chapter on explanation.

Most geological investigations have less clearly quantifiable objectives and the costs of collecting data may not justify detailed simulation of sampling methods. Sampling techniques, however, remain important. It is possible to state in general terms qualities that are desirable in most sampling schemes. The sample should, firstly, be representative of the population, so that valid deductions can be made. Secondly, the sampling pattern should be reproducible, so that from one geologist's description of his sampling scheme, another geologist can collect a comparable set of samples, and use them to test the original conclusions. Thirdly, the sampling scheme should be efficient, so that the effort of data collection yields as much useful information as possible. On the first point, representativity, it is clear that, to give an extreme example, examination of only the largest boulders on a beach would give a biased view of the average grain size of the beach sediment. To give a representative picture, a sample should be collected in such a way that every member of the population has an equal probability of being included in the

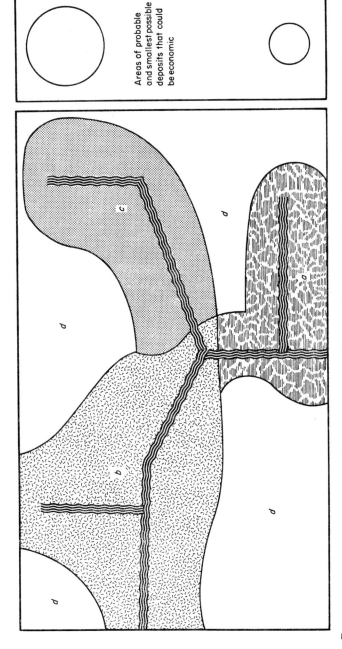

Fig. 17. Graphic simulation of sampling pattern on a map showing river pattern and areas which can be reached by (*a*) boat, (*b*) canoe, (*c*) on foot, (*d*) helicopter. Various sampling schemes can be plotted on the map, and their cost estimated. The risk of missing large and small deposits can be estimated by scanning across the area with the overlay, shown on the right with two round windows of different sizes.

sample. Statistical theory indicates that this can be achieved by adopting a procedure for selecting sample points which is unaffected by and unrelated to any of the properties of the population. In order to achieve the second aim, reproducibility, it must be possible to describe the sampling procedure so that another scientist can check results by collecting a new comparable sample. It is generally neither possible nor desirable to make exactly the same observations as an earlier worker. A change in river level might make an exposure inaccessible, or the original material might have been destroyed during observation. By following the earlier sampling procedure, however, similar patterns should appear when the same population is reinvestigated. The validity of the data can thus be verified independently of any conclusions drawn from them. Where a sampling technique is adequately described, it should also be possible, from theory or experiment, to determine whether conclusions are invalidated by sampling techniques.

Perhaps the easiest technique to describe and visualize is that of grid sampling. It is frequently used in geology. Data collection points are selected on a map of the area of interest by placing on it an overlay with evenly spaced lines running north–south and east–west, see Fig. 18. The position of the grid intersections determines the sampling points and spacing of the

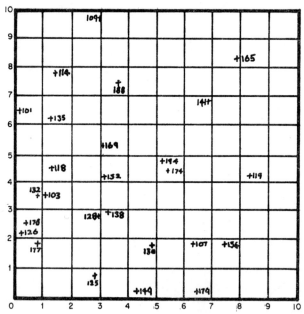

Fig. 18. Random and regular sampling. A full-size transparent overlay with this pattern can be superimposed on a map to select grid sampling points where the lines intersect, or a random sample from the points marked by crosses. In the full-size version, the random accession numbers of the sampling points make it possible to collect a sample of any size.

lines is determined by the size of the sample required. If the position of the overlay on the map is altered, the actual sampling points would change, but the pattern of relationships displayed by the data should not be affected. If the results from several samples differ, it suggests that the size of the sample is too small to reflect the pattern correctly. Grid sampling ensures that data collection is spread evenly across the area. However, it is not suitable if there is regular variation in the property being measured, since this violates the rule that the sampling pattern should be unrelated to the variable being measured. A pattern of evenly spaced sand dunes, for instance, should not be sampled in this way, since the sampling points may tend to fall at a particular point on each dune. Another disadvantage of simple grid sampling is that it gives no information about variation in parameters over a shorter distance than the sampling interval. Samples on a one-kilometer grid would not indicate how much variation would be expected over a distance of one meter.

An alternative technique, which overcomes these drawbacks, is random sampling. Again, a pattern of sampling points can be selected on a map of the area by the use of an overlay. By making a random selection of points, the position of each must be unrelated to the distribution of any feature in the area, and to any other sampling points. A geologist could select random points on a map by throwing darts at it while blindfolded and disorientated. The same result can be achieved less dangerously by rolling dice to select grid references, or more conveniently by consulting a table of random numbers, which are printed as an appendix to some statistics textbooks (see, for example, Dixon and Massey, 1957).

Simple grid and random sampling techniques do not by themselves give a satisfactory approach in a complex geological investigation, and indeed are likely to be less suitable than the haphazard methods which a geologist might adopt intuitively. On the other hand, regular and random sampling can be combined with background geological knowledge to form the basis for more complex methods of practical value. The geologist is frequently not concerned with a simple question, such as 'What is the proportion of sandstone in the sequence?' but rather with a complex model in which it may be necessary, for instance, to collect evidence for an attempted reconstruction of the geological history of an area. The first and most important observations might be directed towards obtaining an overview of the rocks of the area, distinguishing the principal formations and rock-types and mapping their boundaries. Background geological knowledge might indicate that the area consists of formations that are markedly different from one another but reasonably homogeneous internally. The most valuable information is therefore gained by concentrating attention on the boundaries which delineate their extent, and obtaining a comprehensible picture in terms of a conventional geological map.

By such procedures, the geologist can divide the original population into a number of subpopulations, each of which may have different properties and geological significance. Each subpopulation might then be sampled separately, with the largest number of observations, or greatest density of sampling, in the subpopulations of greatest geological interest. Despite variation in the sampling density, however, it might be possible to retain the same sampling pattern throughout. A single criterion is seldom adequate to determine the best sampling procedure. For instance, some areas are more accessible than others. It may be possible to see more rock and collect more information in an hour spent on roadside outcrops than in a week in a remote area of poor exposure. The cost of access must be weighed against the value of the information. It is possible to subdivide a population, such as the rocks of the area under investigation, into subpopulations which reflect ease of access and availability of information. This grouping would almost certainly cut across a subdivision by geological formations. Background geological information can also be taken into account in estimating the value of information from each subpopulation. For example, in a search for uranium ores, it might be known that black shales were a particularly favorable prospect; in a stratigraphic investigation, marine bands or tuffaceous shales might be likely marker horizons, and so on. The trained geologist knows, on the basis of his experience, where maximum effort in an investigation should be placed, and as the work proceeds, can change the emphasis of his observations as he learns more about the area.

Such considerations can still be taken into account when sampling procedures are formalized, and indeed sampling theory may be able to offer guidance in quantified terms about the best sampling strategy. In statistics, the definition and sampling of separate subpopulations is known as multilevel or stratified sampling. The subpopulations are known statistically as strata, but as they have no connection with geological strata, it may be less confusing to refer to them here as subpopulations or levels of sampling. The theory of multilevel sampling brings together various concepts and places them in a quantitative framework. The basic ideas are intuitively obvious. First, by increasing the size of a sample, the number of relevant observations of a population is increased, and more accurate information is gained about it. Second, more precise conclusions can be drawn from a sample of a uniform, homogeneous population than from a highly variable one, and therefore, a larger sample is needed in a variable population to obtain equally precise results. Third, where the population is subdivided into subpopulations, the sampling density of each can be determined separately on the basis of the importance and variability of the subpopulation, and the results subsequently combined to give information about the complete population.

Only the geologist is able to define the objectives of an investigation, the ways in which the population might be broken down for sampling purposes, the estimated importance and variability of each subpopulation, and the likely cost of collecting information in various areas. These aspects together comprise his model for sampling purposes. The model may initially be based on guesswork aided by intuition and previous experience. As the investigation proceeds, it should be possible to refine and improve the model, which may in turn improve the sampling scheme. Repeated sampling may be worth while if an area is readily accessible. An initial sample is collected and analyzed to give information which will guide subsequent detailed sample collection. Data collection continues until analysis shows that the sample is large enough to meet the objectives of the study. A step by step sampling procedure of this type is known as sequential sampling. The cost of reaching an area may however prevent a return to it for more data, and in these circumstances it may be necessary to rely on a predetermined sampling scheme, perhaps modified during collection of the data. The initial stage of the investigation might then be directed towards areas likely to throw most light on the sampling model.

2.8. Properties of a sample from a normal distribution;
sampling design

A quantitative model that has been investigated in detail by statisticians is the normal distribution, considered again in the chapter on explanation. A sample of measurements of a variable which follows the normal distribution has a frequency distribution similar to that illustrated in Figs. 35 and 19. The mean or average value is most frequently encountered, and there are as many measurements above the average value as there are below. Values which deviate considerably from the mean are less likely to be encountered than those close to the mean. The heights of a population of adult human beings, for example, are typically more or less normally distributed. Two parameters that can be calculated to describe a sample of a population are the mean and the variance. The mean, also known as the average, is calculated by adding all the values together and dividing by the number in the sample. The variance is the average squared deviation from the mean, and is a measure of the variability of measurements in the sample.

The mean of a random sample is the best estimate that can be made of the average value of a normally distributed population. It is unlikely to be precisely correct, however, because of the imprecise representation of the population by the sample. Repeated sampling of the same population would give several estimates of the population mean. The variation of the estimates reflects the degree of confidence that can be placed in them. It depends on the amount of variation within the population, and this can be estimated from

F

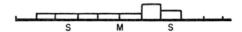

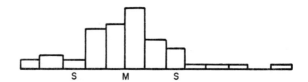

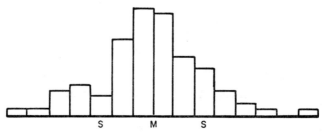

Fig. 19. Histograms of samples cf 10, 50 and 100 items drawn from a normal population. The theoretical normal frequency distribution is shown in Fig. 35. The population mean is marked M, with S one standard deviation (the square-root of the variance) from the mean.

the sample variance. Thus, by collecting a representative sample, and measuring its variance, it is possible to estimate the size of the sample needed to obtain a more precise estimate. Sequential sampling is thus placed on a rigorous quantitative basis. The properties of normal distributions and random samples are known in detail, and are described in many textbooks including Stuart (1962). Many of the conclusions, however, are not confined to strictly normal populations, and the principles of sampling design are widely applicable. The more precisely the criteria are met, the more rigorous are the conclusions, but even where no assumptions of normality can be made about the parent population, the broad principles can still be followed. For example, in the geochemical reconnaissance survey which was introduced earlier, it might be desirable to collect samples of various types, such as stream sediments, rock, soil and vegetation samples. It might be necessary to have these analyzed at various laboratories to obtain an indication of consistency and accuracy. Knowledge of variation in the results is important in making the final interpretation, and must be considered in designing the investigation. For example, if results from soil samples were not directly comparable with rock samples, and if two laboratories were used for analysis, it might be unwise to send all the soil samples to laboratory A and all rock samples to laboratory B, since variation between laboratories could not then be separated from variation between sample types. Before data collection begins, therefore, all recognizable sources of variation should be considered. These might include, in the geochemical example, type of sample, collector, manner of collection, such as depth in soil or position in stream bed, method of sample preparation, method of chemical analysis, laboratory and analyst. With some geochemical equipment, as with human observation, gradual drift away from a calibrated reading may occur, so that results change systematically during a sequence of observations or analyses. Geological variables, such as formation and rock type, will of course affect the results, and weather conditions, such as amount of rainfall, could also be significant.

Having identified possible sources of error and variation, it may be possible to design an investigation in such a way that computer analysis can subsequently disentangle the contribution to the total variation made by individual factors. The usual technique for this is known as analysis of variance, which is mentioned again in the chapter on interpretation. Where many sources of variability can be recognized, the sampling design becomes complex, but the basic principles are a matter of common sense. The pattern of distribution of items among categories affecting one source of variation should not be repeated in another. Rather than submitting all stream sediments to one laboratory, for instance, they might be sent in equal proportion to all the laboratories used. The same procedure could be followed with the other sample types.

If the order of presentation of samples for analysis could affect the result, because of instrumental drift or because the analyst directs successive batches of samples to different apparatus, then the samples should not be presented in the order in which they were collected. Otherwise, an apparent anomaly or regional trend might appear in the final maps, which was in fact due solely to instrumental errors. It may be possible to split a sample into two equal parts, or to collect two samples from the same location, and to have these analyzed separately at different times or on different instruments. If a number of samples are duplicated or replicated in this manner, the results can be tested to determine whether they are consistent. Where sources of error can be estimated and allowed for by computer, it may be possible to obtain satisfactory results from less reliable, but cheaper analytical methods.

Geochemical surveys are an obvious example of investigations where the geologist can, by careful design, identify and control sources of variation and error in the survey (see, for example, Plant *et al.*, 1975). The same is true in many other fields, such as geophysics, petrology, hydrogeology or engineering geology. The sources of variation can be controlled because they are introduced by the method of investigation. In many field studies, however, the processes which caused the most important variation were complete many millions of years ago, and the product is only partly accessible to observation.

The objective of sampling design in such a situation must be to establish which variables and relationships are important and how their properties may most efficiently be established. It may be necessary, for example, to determine whether data collected from rocks accessible at outcrop can also be considered representative of the unexposed strata in the area. A comparison of well exposed and poorly exposed areas might throw light on this point, if the latter are more likely to resemble the unexposed rocks. Consideration of the outcrop pattern, and the reason for the position of the exposures, might indicate whether it was likely to be related to pattern of variation in the strata. If not, then it might be possible to collect a representative sample at outcrop. But if a relationship was suspected, either it could be investigated and allowed for, or the conclusions drawn from the sample could be appropriately restricted. A case might be made for taking shallow borings in areas of no exposure if the value of the information was likely to outweigh the expense of collecting it.

In a field study, the objective is not usually to test one simple hypothesis, but rather to investigate a complex model, in which many relationships are of interest, and many variables can be linked to a single cause. A sampling scheme based on areal distribution can be totally inadequate in these circumstances. For example, if in a study of structural geology, it is important to know whether folds are generally cylindrical or conical in form, then all the orientations of beds around the folds should be equally represented. If, as is not unusual, the folds have sharp crests and long limbs, as in Fig. 20, the orientations of the limbs would be over-represented in a grid sample. The geologist might therefore define various classes of orientation of bedding, as in Fig. 20, and attempt to collect a random sample within each class, so that all are equally represented. Such a sample might provide the desired information about fold shape, but could not be used directly to obtain a value of the average orientation of the beds. The concept of average orientation is tied to area, and to calculate it correctly, all parts of the area must be equally represented. This could have been achieved by taking a grid sample. Alternatively, a weighting procedure could be used, as described below.

An analogous situation could arise in the example mentioned in introducing the topic of sampling, namely, an investigation of geochemical changes near

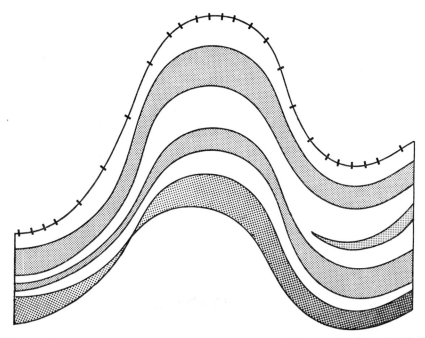

Fig. 20. To determine the axis of folding, and whether the folds are conical or cylindrical the orientations of bedding planes may be more densely sampled near the crests of folds, ensuring that all orientations have an equal chance of being sampled. Sampling points are indicated by bars.

the margin of a granite intrusion. If, for instance, relations between iron and magnesium content were of interest, and if this changed rapidly at the granite margin, then in order to obtain a sample which represented the relationship, the density of sampling might have to be greatest at the boundary of the intrusion, decreasing with distance from the boundary. Again, it might be possible to define a set of subpopulations as in Fig. 21, and sample randomly from each. To obtain average values for the iron and magnesium concentration throughout the area, and separately for the granite and the country rock, would require a sample linked to area. With two objectives calling for different sampling patterns, it might be desirable to collect the two sets of samples independently. However, the cost of performing a geochemical analysis can be high compared to the cost of collecting a sample, and the effort of designing a sampling scheme to serve both purposes could well be justified.

Taking the example in Fig. 21, equal numbers of points might be sampled randomly from each of the subpopulations marked. The boundaries of the subpopulations are chosen to give a reasonably even representation of the

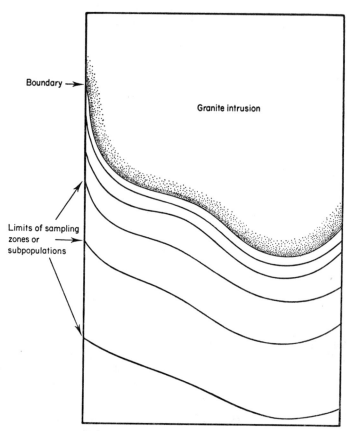

Fig. 21. Sampling pattern near a granite boundary. To investigate the rapid geochemical changes near the boundary, the same number of samples is taken in each of a set of successively narrower sampling zones.

range of values of iron and magnesium in the granite and in the country rock. It would not, presumably, be possible to define the subpopulation boundaries precisely before the study was complete. But previous experience or a preliminary examination might suggest a working model; perhaps that the rate of change of chemical composition decreased exponentially in both directions away from the margin. The subpopulation boundaries could thus be selected at distances on a logarithmic scale from the granite contact, and parallel to it. When hypotheses are tested which involve area, such as the hypothesis that on average the granite contains more iron than the country rock, then the various subpopulations can be weighted on an areal basis. As shown in Fig. 21, each successive subpopulation covers roughly twice the area of the preceding one. To obtain an average value for the granite as a whole, therefore, the subpopulation averages are added together in proportions depending on their area. Analysis of this type of data is an example of a general principle, of weighting subpopulations according to their significance in the model.

A geological field investigation is seldom so simple or clear-cut that the models under investigation can be well defined in advance. In practice, several sets of cross-cutting subpopulations might be involved. The aim is to sample the population in such a manner that as much information as possible is collected which is of direct relevance to the models under investigation. This may impose a strong directional element in the sampling scheme. For example, in a strongly folded area, there may be rapid changes in bed orientation in a direction at right angles to the strike, whereas changes in a direction parallel to the fold axes might consist merely of gentle culminations and depressions along the folds. In a similar way, beach sediment may change rapidly across the beach from wind-blown dunes to storm beach deposits to coarse gravel at high-water mark, grading to medium-grained sand below low-water level. Parallel to the shore-line, on the other hand, the sediment may be comparatively uniform in its properties, changing gradually over a distance of several miles. The sampling scheme can take these directional properties into account. A distorted rectangular grid might be an appropriate sampling pattern in the above example, with measurements at, say, ten yard intervals on traverses across the beach, and with the traverses spaced perhaps 500 yards apart. In the case of the folded rocks, strike and dip might be measured at approximately 100 yard intervals, say, on traverses across the fold axes, with occasional traverses at right angles to investigate changes in that direction. Convenience of access will inevitably dominate most field sampling, but need not affect its representative nature, provided the sampling pattern which it controls is either not linked to any relationship in the model under investigation, or is related in a known way which can be taken into account in drawing conclusions. It must be possible to describe the sampling pattern if other geologists are to use the raw data, so that they may be clearly aware of its limitations and the extent to which they can draw valid conclusions of their own from it.

Two final points are perhaps worth emphasizing as having a bearing on the sampling procedures adopted in an investigation involving computers. The first point is that numerous characteristics of the rock are generally investigated. Only by recording them systematically can they be analyzed systematically. There is a vital distinction between the fact that a property, say cleavage, was not looked for at a particular point, and the fact that it was looked for but not found. It is a distinction, however, that is not always clear in field records. If, as is usually the case, several properties are recorded, they should be observed as far as possible at the same place or on the same specimen. Thus if the petrology and geochemistry of an area are both being investigated, then as far as possible the same specimens should be analyzed both petrographically and geochemically. Only then can like be compared with like, and the links between the two aspects established. The second point is that if valuable information is observed it should be recorded whether or not it fits a preconceived sampling scheme. It is always possible to label such observations as being non-typical, and omit them from those parts of the computer analysis where they might be misleading. A large part of any geological investigation is likely to consist of looking for and attempting to explain the unexpected. Theoretical concepts should never outweigh common sense. The geologist's own appraisal of his subject matter is all-important.

REPRESENTATION OF DATA FOR THE COMPUTER

2.9. Constraints imposed by the computer

Geological data are a factual record of observations. The geologist determines the structure and content of his data taking into account the system he is describing; established conventions and his own previous experience; his objectives in the investigation; and the ways in which he plans to analyze, present and interpret his results. Where computer methods are introduced, they may bring requirements for new conventions and new methods of analyzing and interpreting the data. On the one hand, computer methods impose constraints on data recording; on the other, they offer opportunities for more rigorous statistical analysis and wider data exchange. The constraints are worth considering in more detail, since they affect decisions on the extent of computer involvement and the mode of data collection.

Clearly, unless adequate resources will be available to process the data, there is little point in collecting them. The resources include hardware access and availability, sufficient expert knowledge to use or to develop the appropriate techniques, and suitable software and systems. The cost of preparing data in machine-readable form must be borne in mind, as this may account for more than a third of the total expense of the computer activity. In a large-scale investigation, even the cost of transcribing data to coding sheets could be a major item.

Computer languages impose further constraints on data recording. Fuzzy, imprecise data sets may greatly assist the geologist to develop his interpretation, perhaps using such modes of thought as analogy and intuition. But they do not lend themselves readily to precise, unambiguous digital representation. Statements in a computer language represent simple mathematical or logical operations, and facilities have to be limited, so that the language is reasonably easy to learn and use, and is capable of producing efficient programs. Because each step in computer analysis must be explicitly set out in the program, the format and content of the data files must be carefully disciplined. Otherwise, the need to allow for all exceptions to the general pattern would result in programs that took too long to prepare, were difficult to maintain and expensive to run. Data for the computer, therefore, should normally be recorded in a uniform manner for each item. As mentioned in the section on software, FORTRAN and similar languages differentiate between real, integer and text variables, and store them in different internal representations. It is thus generally undesirable for computer purposes to mix data types in one variable, for example, by recording the dip as 'gentle' at one location, and 15° at another. Instead, either quantitative estimates could be recorded throughout, or else a set of categories could be used consistently which would give a useful breakdown of the results. Of the wide variety of data structures in which geological data can be recorded, it is only the array that can be handled naturally and simply in some widely used languages such as FORTRAN.

In general, therefore, it is advisable to decide on the format and content of data at an early stage in a computer investigation. If the data are to be processed by FORTRAN programs, then the values of one variable will be handled as all text, all real or all integer. It is usually sensible to record data for one variable to roughly the same precision in all records. If the average content of silica is to be determined for a group of rocks, and is recorded as about 50% in one and 20% in another, there is little point in recording 25·231% in a third. It may help if the decimal point is indicated in each field of the coding form. The decimal point can occupy a column on the form and is then actually punched on the card, or its position can be shown between two columns, if its location is known to the computer program through the format statement. If, for some exceptional reason, the precision differs between items, the user can be instructed to insert the point when recording the data. Numerical data are normally right-adjusted within the field, so that digits of the same magnitude appear in the same vertical position in the form. Text data, on the other hand, are conventionally left-adjusted so that they can be sorted into alphabetical order. In Fig. 22, a small triangle indicates whether each field is left or right adjusted. To avoid confusion, the figures 1 and 0 can be written as 1 and $\emptyset$ to distinguish them from the letters I and O, and care should be taken in writing Z and 2. Any conventions of this kind, including the symbol for a blank space, should be brought to the attention of the data preparation operator, perhaps by a note at the head of the data sheet.

The characters in any field that is to be used for sorting must be positioned correctly within the field. When characters, including the space or blank

I.G.S. SUBSURFACE LAND RECORDS Summary Borehole Succession

Registration number: 6 in. quarter sht. NS98NE (1–6) | Accn. no. 119 (10) | Suffix (14)

Header record: D (15) | Author of strat. WAR (19) | Year 1963 (23) | Other strat. classns. N (24, Y or N) | Imperial or Metric depths I (25, I or M)

Signed: W A Read | Date: 15/3/76 | Page 1 of 1

Succession records:

15	ACCN NO. OF ROCK UNIT (16–18)	STRAT. CODE: LITHO-STRAT. (19–22)	STRAT. CODE: CHRONO-STRAT. Lwr (23–24)	CHRONO-STRAT. Upr (25–26)	RELIABILITY (27)	DEPTH TO BASE: FEET / METRES (28–32)	INS / DEC. METRES (33–34)	1/10 INS (35)	LITHOLOGY: 1st COMPONENT (36–39)	RELATION (40)	2nd COMPONENT (41–44)	RELATION (45)	3rd COMPONENT (46–49)	COMMENT (50)	BASE OF BED (51–52)	INTERVAL RELIABILITY (53)	STRATIGRAPHY (Clear text/diagrammatic)		
E	1		QF			1	ø2		PEAT									HOLOCENE	QUATERNARY
	2		QP		1	4	ø9		CLAY		SAND			*	ES			?PLEISTOCENE	”
	3	PGP	CG	CY	3	7ø6	ø9		SDST		SLST		MDST	*	ES	1	Top not cut / PASSAGE GROUP / Disconformity	NAMURIAN	UPPER CARBONIFEROUS
	4	ULG	CG		ø	1372	ø6		SDST		SLST		MDST	*		2	UPPER LIMESTONE GROUP / Base not cut	”	”
	5	END																	

LDB/F/2A.4

Fig. 22. Data sheet for mixed quantitative and descriptive data. Reproduced with permission from Farmer and Read (1976).

(denoted ⊔), are stored in the computer, the internal representation is such that when the binary coded characters are arranged in ascending order, the characters they represent are in alphabetical order, generally with ⊔ preceding the digits 0–9, preceding the letters A–Z, preceding the special symbols such as punctuation marks. After sorting into ascending sequence, therefore, the following items of four characters: ⊔ZZZ;   1ABC;   AAA⊔; ABC1;   ABC1;   Z⊔⊔Z; Z⊔A⊔ and ·234 would be arranged in that order. The placing of blanks or spaces is significant and care is needed to ensure that they are placed consistently. The position of blanks before or within a word must be clearly indicated. For this reason, forms for data recording frequently have boxes to indicate the position or each character as in Fig. 22. Alternatively, the beginning or end of each field may be clearly marked and an indication given of whether the field is left- or right-adjusted. Spaces within the word, sometimes known as embedded blanks, are then indicated by a special symbol, such as ⊔ or ∇ or ƀ. This is an indication to the keypunch operator to press the space bar once. Since a card punch or data entry terminal can be set to skip to specified columns, blanks following a word need not be indicated if they are to be punched on cards with a fixed field length. The operator presses 'skip' on completion of each field, and the card moves automatically to the beginning of the next field. Much greater freedom of layout is provided by systems for free-format text handling (see Forbes *et al.*, 1971) as is illustrated by the computer-input form of Fig. 23.

The majority of geologists who use the computer do so through existing programs or software systems. For many applications, the choice of program determines the format. The method of analysis also strongly influences the structure and content of the data. Many methods of multivariate analysis, for example, require data as a quantitative matrix. In general, computer methods call for systematic and consistent techniques of data collection and recording and advance planning is needed to achieve this. In a complex investigation, detailed systems analysis will be required at an early stage to define the data files, and their arrangement, format and content.

2.10. Potential uses of computer data

In order to guide data collection, the potential uses to be made of data files must be considered. A broad classification, from the viewpoint of the geological systems analyst, might be the following:

A data with no computer involvement;

B data indexed on the computer;

C text commentary stored on the computer;

DEPARTMENT OF GEOLOGY   CAMBRIDGE

LOCALITY CATALOGUE   SET CSE   NUMBER G 882

LOCATION	PLACE Cambellryggen	DISTRICT Bunsow Land
	COUNTRY Spitsbergen REALM Arctic	HEIGHT

LAT/LONG	° ' "	° ' "	LONG ORIGIN	° ' "

GRID REF	SYSTEM UTM SQ 33X/VH E 140 N 229

DETAIL	Northwest spur above Brucebyen

TYPES OF DATA COLLECTED	SECTION	BOREHOLE	RADIOMETRIC	PALAEOMAGNETIC
	PLANT	INVERTEBRATE	VERTEBRATE	PALYNOLOGICAL
	MICROFOSSIL	TECTONIC	LITHOLOGICAL	SED STRUCT
	TOPOGRAPHIC	PHOTOGRAPHIC		

ROCK UNITS	Nordenskioldbreen Formation, Minkinfjellet Member, Cadellfjellet Member, Tyrellfjellet Member, Brucebyen Beds.

ROCK AGE OR COMPLEX	Moscovian, Gshelian, Asselian

SPECIMEN NUMBERS	G1130 to G1170

UNPUBLISHED DOCUMENTS	TYPE REFERENCE	DATE	PLACE
	MS	1959	CSE files

PUBLISHED REFERENCES	

VISITS	BY Gobbett, D.J.	DATE 1959

NOTES	

CATALOGUER	DATE S.M.2 MAY 1969

Fig. 23. Data sheet for recording free-format data for the computer. Reproduced with permission from Forbes *et al.* (1971).

D data for translation and conversion;

E data from which simple summary statistics are to be calculated;

F data for mapping or computer display;

G data for detailed numerical or statistical analysis by computer.

Information which will be interpreted and used entirely without computer assistance (A), such as text descriptions, maps, analog charts and diagrams, can be held in conventional form without any need to consider computer formats. The simplest form of computerization is preparation of a computer index (B). For this, each document in the file must be uniquely identified, perhaps by numbering each in sequence with an accession number, and a set of retrieval codes must be selected to describe important aspects of the document for retrieval purposes. The result of computer search is then a list of reference numbers which in turn make it possible to find the original document. It may be convenient to hold the original records on microfilm or microfiche to simplify the physical task of finding and possibly copying and mailing the documents with the known identification numbers. The cost of storage and reproduction can be greatly reduced by this means. In some applications, facsimile transmission is now an economic proposition for making computer-indexed material held at a central location available at a number of points.

It may be desirable to store data in a computer solely for retrieval and presentation (C). The motive is generally to have the data linked to other computer information. For example, if depths to formations in a set of boreholes are stored in a file for full computer analysis, it might be desirable to have related comments on the reliability of the identification, perhaps with the name of the stratigrapher, date and basis of identification also held in the computer so that they can be retrieved and printed out as comments whenever the stratigraphic data are analyzed. The main data and the comments might be held in separate subfiles, as they have different structures and formats and are processed in a different manner (see Fig. 24). This is similar to a manual system where index cards carry limited information in predetermined positions, and are cross-referenced to records of varying size and type in a filing cabinet. In a computer system, the index information might be held in a two-dimensional array, with each row cross-referenced to a string of data of variable length.

A slightly more involved application might call for the printing of information rearranged in a number of different ways. For example, listings of a set of comments records might be required, printed first in order of occurrence, second by formation name, third by borehole location. It would

I.G.S. SUBSURFACE LAND RECORDS Important Comments on Summary Borehole Succession

Registration number	6 in. quarter sht	Accn. no.	Suffix	Record type
N S 9 8 N E		1 1 1 9		F
1 6		10	14	15

Author of stratigraphy to which comments refer	W. A. READ
Year of this strat. succession	1963

Comments Page 1 of 1

→ Comment records in succession order

ACCN. NO. OF ROCK UNIT

COMMENTS (referring to specified rock unit)
Use block capitals and as many lines as required (up to 10)
Start comments for each unit on a new line

```
 1  POST POLLEN ZONE III
 2  EXACT AGE AND ORIGIN UNKNOWN
 3  ALSO THIN LMSTS AND COALS. MOST
    MDST IS FIRECLAY. SOME SDST IS
    ILICA ROCK (REFRACTORY). UPWARD-
    FINING FLUVIAL CYCLES. TOP OF GR
    OUP NOT CUT. BASE DISCONFORMABLE
 4  ALSO LMSTS AND COALS (SOME WORKA
    BLE). MOSTLY DISTAL DELTAIC. BAS
    E OF GROUP NOT CUT
 5  END
```

→ Comment records continued in succession order

ACCN. NO. OF ROCK UNIT

COMMENTS (referring to specified rock unit)
Use block capitals and as many lines as required (up to 10)
Start comments for each unit on a new line

LDB/F/2B.4

Fig. 24. Data sheet for variable length comments, cross-referenced to the succession data of Fig. 22. Reproduced with permission from Farmer and Read (1976).

not be necessary to sort the comments file for this, since all the rearrangements could be made on the index file, just as the index cards of a manual system can be rearranged without altering the detailed information in another filing cabinet to which they refer. This is therefore another reason for creating distinct subfiles for data which are to be stored and handled in different ways. Text comments are one example of data held as variable length strings. Others include sets of replicate measurements of a parameter and sets of points defining a line on a map or a seismic record. The mode of representation of each subfile can be linked to its possible usage.

Conversion from one code to another (application type D) can be performed on the computer by table lookup as in Fig. 25. For example, short mnemonic codes may be convenient for recording lithology or stratigraphic horizon, being comparatively easy for the geologist to remember, but unsuitable for computer processing, where it could be desirable for formations to be numbered in stratigraphic sequence for ease of retrieval and sorting. If conversion is by table lookup, then the codes used on the input documents must have precisely the form of those in the lookup table. Careful

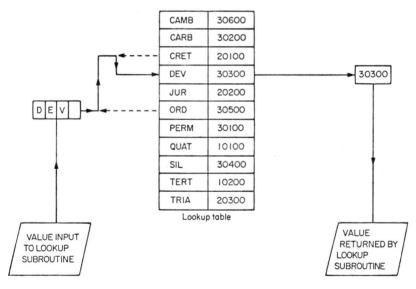

Fig. 25. Table lookup. The mnemonic code DEV has been read by the computer as data, and the lookup subroutine attempts to find the same code in the first column of the lookup table. One strategy with codes arranged in alphabetical order, is to make the first comparison with the central value to determine whether the code lies above or below. The next comparison is with the central value of the remaining range of possibilities, and so on until a match is obtained. The numeric code in the second column is returned to the program by the subroutine.

form design and the use of an up-to-date checklist of acceptable codes by the geologist may be necessary to ensure this.

An example of the calculation of summary statistics (E) would arise where a paleontologist recorded the occurrence of the main fossil species at several localities. The type of statistics to be calculated might include: number of beds containing each species; percentage of occurrences of Lingula at which the presence of plant fragments was also noted; or the average thickness of beds containing plant fragments. Data of a similar type might be involved in computer display (F) by plotting occurrence of species against depth, or a locality map. Detailed numerical and statistical analyses (G) are perhaps the applications most widely associated with computers. Examples are the heavy calculations involved in correcting and interpreting seismic records, or investigating by multivariate analysis the relationships between the components in a set of geochemical analyses.

2.11. Files and subfiles; data sheet design

The first step in considering data collection in an investigation is to attempt a broad subdivision of the data into files. A file constitutes a group of similar data records on one topic, such as geochemical analyses from Map Sheet 121 of the Ercu Mountains. For convenience, a file may be broken down into separate subfiles of uniform data content, containing, for example, major chemical elements determined by wet chemical methods, elements determined by X-ray analysis, and elements determined by optical spectrometer. In general, computer files might be organized in such a way that there is maximum connection between variables or items within a file, and as little requirement as possible to analyze the information from a number of files together. Subfiles, on the other hand, might be selected to ensure that, as far as possible, all records within the subfile have the same data content. The information on a data sheet would generally refer to a single subfile. Information about one object might appear in several subfiles. For example, data concerning one stratigraphic unit in a borehole might appear in subfiles concerned with lithology, stratigraphy, geochemistry and paleontology. The same unique key, in this case borehole and bed number, should then appear in each subfile and on the corresponding data sheets, to ensure that cross-references can be established.

In complex studies, there are many distinct sources of information which give rise to related data files. Each file can be designed separately, but overlap in the contents of different files must be examined to avoid unnecessary duplication in data collection, and to ensure that there is consistency between sources of information that have a bearing on the same geological conclusions. At an early stage in the systems analysis, it is desirable to list all the known data files and to consider their relationships. The decision as to what constitutes a data file is a somewhat arbitrary one. As a first attempt, a broad subdivision into large files may be appropriate, leading to finer subdivision at a later stage in the analysis.

The extent of the computer involvement in each file depends on many considerations, in particular, the benefits to be gained from computer analysis or data management, as well as such aspects as the scope of the data, in terms of the number of individuals collecting data in the investigation, the number of potential users of the data, and the variety of uses to which the data would be put. The frequency with which users are likely to access the data, and the length of time before the data become redundant or obsolete also have to be considered in deciding how best to handle the data.

In some cases, data can be recorded in more or less the form in which they will be prepared for computer input. For example, data sheets can be completed in the field that are later passed directly to a keypunch operator. As well as saving duplication of effort, this approach ensures that the originator of the data has a clear idea of what will be presented to the computer. It is not an appropriate procedure, however, if the uses of the data are not known in advance, nor if little computer processing will be involved. It may then be preferable to record the data in a more conventional manner, and transcribe parts of them as necessary for computer input as a separate and later activity. There are advantages in holding the primary or master record on the computer, but, since the field of geological computer applications is still relatively new, it is safer to at least maintain a duplicate master file which can be used without computer assistance, as a safeguard in case the new methods fall short of their expected performance. It may be possible to consider the original data forms as the duplicate master file provided the information is reasonably comprehensive and not too highly encoded.

At an early stage in the systems analysis, it may be helpful to prepare rough drafts of forms or data sheets such as those illustrated, to assist in visualizing the data content and in planning the data collection procedures. But the final form design should wait until the draft forms have been tried in practice. Many points have to be borne in mind in setting data collection procedures. Geological considerations determine the content of each file, but careful analysis is necessary to ensure that relevant, consistent, complete and unambiguous data are obtained in an efficient and flexible manner. Traditional methods of data collection are not always entirely successful in these respects, and systems analysis can lead to better methods even where computers are not employed. Relevance of the data to the objectives of the investigation is one of the most important aspects, and one of the most difficult to achieve, since the objectives themselves may change as data are collected and the problems are better understood.

G

Flexibility is thus another aim in data collection, sometimes in conflict with the aim of consistency. An acceptable compromise may be to ensure that the data collection procedures do not inhibit the recording of unforeseen information, and that the procedures are regularly reviewed and brought up to date. Any revisions should be documented so that it is known, for instance, whether data recorded on a certain date had no record of sedimentary structures because none were seen, or because none were being looked for. The form in Fig. 22, supplemented by check lists, is intended to encourage consistency in recording the main items of information, while giving scope to add unforeseen items in the comments section (Fig. 24). Major items of additional data, such as geochemical analyses, can be incorporated in separate files, linked by an entry in the cross-reference field. Other practical aspects to the important question of form design, which is a major topic in itself, are considered by Daniels and Yeates (1969).

2.12. Fields and records

Where the mode of representing the data within the computer is not already established by existing conventions, the geologist may have to consider the alternatives more fully, either from the point of view of selecting appropriate software or of preparing his own. It is in any case desirable to understand in outline the various possible ways of presenting data in the computer. The media such as punched cards, disks and magnetic tapes, on which data can be stored, were mentioned in the section on hardware. Bit patterns were described in which, for example, eight bits (or a byte), or one column on a punched card, represent a character. Several bytes, generally four, comprise one computer word. The word (generally 32 bits) can also hold a quantitative value in integer or floating point form. A set of adjacent characters or numbers which represent one name or value, and are intended for processing together, is known as a field. A set of fields that are to be considered together, such as a set of attributes and measurements referring to one sample, are known as a record, while a collection of similar records on one topic constitute a file or sub-file.

2.13. Input, exchange and storage formats

The arrangement of data prepared for computer input is known as the input format. When the data are read by computer, the various bit patterns are translated into another form for internal representation, and bytes, words and records may be rearranged to some extent. Similar conversion processes occur when the data are stored on disk or tape, or when they are manipulated by a processing program, written to a tape or communications device for transfer to another computer, or transmitted to an output device for printing or plotting. There are thus several formats in which the data are held at various times, including the input format, the storage format, the processing format, the exchange or communications format and the output format. The internal representations and the translation between them are controlled by the operating system and need not concern the user. However, the formats can be manipulated by the user through his programs, and it is important to remember that different formats can be selected for different purposes. Thus, data can be communicated between two systems using a common exchange format without affecting the mode of input or storage in either system. As usual, a trade-off is involved, in this case flexibility being gained at the expense of more complex programs and more extensive processing. The exchange format can make it easier to exchange data among several systems, since each system need interface with only one external format.

The input data file must carry enough information to enable the computer to organize the fields correctly for processing. The position of the fields on the input record is conventionally indicated in FORTRAN by a format statement, described in the software section. Each field occurs between the same columns on every input record, and is thus always in the same position in the sequence of fields. The format is fixed and there is no need to name the field in every record. An alternative form of representation is free format, in which a field is not necessarily of the same length in each record, and indeed, is not necessarily in the same position in the sequence of fields. The field boundaries in free format are indicated by a separator or delimiter. This is a character or sequence of characters, such as a comma, or one or more blanks, which are designated as separators, and cannot therefore be used within a field. An advantage of free format is that data values of any length can be accommodated, and the maximum length need not be decided in advance. On the other hand, for data that can readily be represented in the form of a table, fixed format gives a straight-forward easily checked arrangement of input data, which can be handled directly by a FORTRAN program. For tabular data, it is also convenient to have each field occupy a constant position in the record. However, where a field occurs in only a small proportion of records, it may be more efficient to record both the field name and value where they occur, rather than storing an empty field where they do not. If the fields and field sequence vary from record to record, each value can be preceded by the field name, so that field names and values alternate across the record. These are sometimes known as tag/value or label/attribute pairs. With ingenuity, many variations are possible, such as having a rigid sequence of fixed format data, followed by an optional number of label/attribute pairs, with different separators (say, and.) to indicate the end of field and end of record. Programs can be written, even in FORTRAN, to handle such data, but the complexity increases the difficulties of programming and the possibilities of error. Techniques for analyzing tabular data are well established, and the benefits of the proposed methods of analysis must be carefully considered, to justify mixed and more involved structures outside a research context.

Trees and networks as data structures were mentioned in the previous chapter. The use of tag/value pairs is a convenient means of representing a tree structure in input format. In storage or processing format, however, it is more usual to use pointers, which, as illustrated in Fig. 26, indicate the address in the computer store of the next relevant occurrence of a variable or of successive items in a record. That occurrence is then linked with a pointer to the next occurrence again, making it possible to trace a path through successive values of a variable. Branches in the tree are indicated by replacing the value/pointer pair by two pointers which indicate the position of the next values, one on each branch. Additional refinements, such as associating a pair of pointers with each value, indicating the next and previous values, make it possible to follow the branches of a tree in either direction. Yet another pointer can be added at more widely spaced nodes, linked to one another, so that an uninteresting part of the path can be skipped over rapidly without stopping at every value. With more complex data, sets of pointers can be used to associate each data item with other information, such as variable name, units of measurement, or form of relationship.

Pointers are one means of indicating cross-references between data items. Links between associated data items can also be established by ensuring that entries in key fields, such as location, stratigraphic unit, or well number, are uniform in all relevant files. It is then possible by selective retrieval to recover all data referring to a particular outcrop, formation or borehole. As it involves extensive searching, this is unlikely to provide the most efficient means of cross-referencing, although the difficulties can be greatly eased by declaring certain fields to be specifically for cross-referencing, and ensuring that terms are used consistently in these fields at least. Where cross-linkage at a relatively simple level is required between records in two or more files or subfiles, for example in linking records for the same outcrop in files of paleontological, petrological, lithological and stratigraphical data, an additional cross-reference file can be created which holds the record numbers of data for each outcrop in turn, in each of the files (see Fig. 27). A change in record number in any file would then require that the corresponding amendment be made to the cross-reference file. Another kind of cross-reference file is known as an inverted file. It consists of the entries in a key field, sorted in alphabetic order, together with the address of the

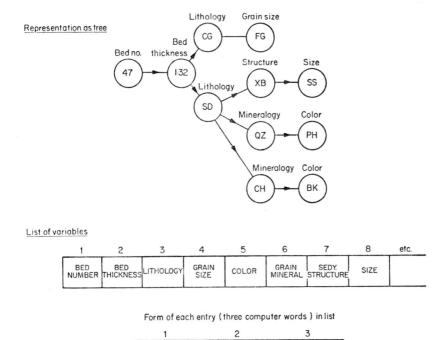

List of variables

	1	2	3	4	5	6	7	8	etc.
	BED NUMBER	BED THICKNESS	LITHOLOGY	GRAIN SIZE	COLOR	GRAIN MINERAL	SEDY STRUCTURE	SIZE	

Form of each entry (three computer words) in list

1	2	3
DATA ITEM	POINTER TO VARIABLE NUMER	POINTER TO NEXT ITEM

At a branch, the data item is blank and the next two words are the addresses of the next two data items

List representation of data

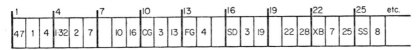

Fig. 26. The representation of a tree-structured description by a list with pointers. Lithological description: Bed no. 47, 1·32 m thick. Sandstone; small-scale cross-bedding; grains of quartz, purplish and chert, black. Conglomerate, fine-grained.

full record in the original file. It serves the same function as the index to a book, in providing an efficient means of locating a record when the value of the keyfield is known. These housekeeping tasks are handled automatically within many data management systems, and indeed cross-reference between files becomes much more difficult if they are not all held within the same software system. Data management systems are considered later in this chapter, following consideration of the mode of representation of entries in individual data fields, and the use of codes of various kinds to make this possible.

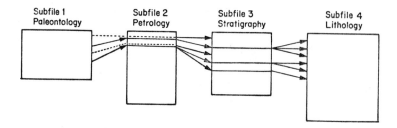

| Subfile 1
Paleontology | Subfile 2
Petrology | Subfile 3
Stratigraphy | Subfile 4
Lithology |

	RECORD NUMBER IN SUBFILE			
OUTCROP NUMBER	SUBFILE 1	SUBFILE 2	SUBFILE 3	SUBFILE 4
1	1		1	
2	2	1	2	1
2	2	1	2	2
2	2	1	2	3
3	3		3	
4	4	2	4	4
4	4	2	4	5
4	4	2	5	6

Fig. 27. Cross-reference file. The file contains the addresses of records in the various sub-files which refer to the same item.

2.14. Coding and abbreviation

A data value may be either quantitative and represented as a binary code in the computer, or held as a string of text characters for text handling rather than numerical analysis. Information about a rock or mineral type, a fossil name, or a stratigraphic unit is likely to be in the form of a text string. There are a number of ways in which these can be recorded and stored in the computer, and, as before, the input, storage, processing and output format need not be the same. As an example, information might be obtained on the fossils collected at several localities, together with related information such as the lithology, thickness and position in the bed in which they were found, their abundance, state of preservation, etc. This information might conveniently be organized in a tabular format, in which one column carried the fossil names. There are disadvantages in recording the name in full. One is that the column would have to be wide enough to accommodate the longest species name. Space would thus be wasted on the input form and in the computer. A second drawback is that time would be wasted by both the geologist and the keypunch operator in writing the name in full. A third is that consistent spelling of names may not be easy for either the paleontologist or the keypuncher, and checking the spelling is a laborious task. Fossil names can include symbols such as ?, aff. and cf. in various positions and this causes difficulty in sorting the names into alphabetical order. Finally, the full names do not help in arranging the species automatically in biological groupings, or by stratigraphical or environmental significance.

Faced with these difficulties, the systems analyst might decide to represent the fossil names by codes, remembering that input, storage, processing, exchange and output codes need not be the same. A simple input code could be obtained by assigning letters or numbers arbitrarily to the various fossil types. The real name of the fossil might even be considered to be irrelevant, and codes assigned on the basis of the characteristics of the fossil. To ensure that codes are not accidentally repeated, accession numbers can be used, with successive numbers assigned as required. Telephone numbers are codes of this type. With many different fossil species, however, it becomes difficult to remember the codes accurately. It may be helpful to use similar numbers for similar fossil types for this reason, with the additional advantage that, when sorted by code, related species are brought together. This approach is used in the UDC and Dewey Decimal systems for book classification, where one important objective is to ensure that books on the same topic are brought together on the library shelves when they are arranged by code number. This system could at first sight give a good stratigraphic code by assigning successive numbers to formations arranged in stratigraphical order. The difficulty is that an array of successive numbers is 'inhospitable', that is, it cannot readily accept insertions of new numbers. With stratigraphic formations, as with subjects of books, it is likely that new formations will be added, and, perhaps because of exploration of a new area, several new formations may be recognized in the same part of the stratigraphic column. By numbering in tens, space is left for inserting later additions, thus: 10, 20, 30, 40 . . . But, having added a new formation as 81, to follow 80, there is no room for a new formation between 80 and 81. The decimal system of numbering solves this problem by placing a decimal point after the original codes and adding more numbers, and more places after the decimal point, as required. The resulting rather long and cumbersome numbers can be seen in any library, particularly on the spines of books of fast growing subject areas like mathematical geology. A more fundamental difficulty with subject-related codes is a possible ambiguity of classification. Should mathematical geology be regarded as being closer to mathematics, or to geology? Should the code for granite be similar to that of gabbro, because of grain size, or rhyolite, because of composition? A code which reflects similarities and relationships implies a classification, and will reflect any deficiences in that classification.

It is rather easy to introduce an error in using a simple code, such as accession number, perhaps by accidental transposition of digits during initial recording, transcription or keying. Thus, 1697 might be written as 1967. With a numeric code, transposition can produce another code which, although it has a totally different meaning, is still valid, and might thus go undetected. One means of preventing this is to add an additional 'check' digit to the code, producing an 'error-checking code' on which a simple arithmetic check can be carried out to determine whether or not the code is valid. For example, a digit could be added to the end of each code to make it exactly divisible by 11. Thus 1697 would become 16973. Transposition, changing or dropping a number is then likely to produce an invalid code which is no longer exactly divisible by 11 and can therefore be readily detected by a simple computer program. In a similar way, check characters can be added to alphanumeric codes, perhaps calculated from the binary representation of the individual characters.

Numeric codes are not easy to remember, as shown by the relative difficulties of remembering a colleague's name and telephone number, and using an extensive set of codes requires frequent access to a directory. Mnemonic codes, which are designed to have the quality that they are easy to remember, are therefore frequently used in data preparation and sometimes in output. The usual reason for using them is to avoid unnecessary writing and keypunching. However, unlike a simple numeric code, transposition errors tend to be easy to find and correct. Compare, for example, the typing errors 'see page 123' and 'see paeg 132' where 'see page 132' was intended. It is usual for the geologist to employ contractions and abbreviations as mnemonic codes in the initial recording of data in field maps and notebooks, with such notes as: 2'3" sndst, lt gy-brn, f-grnd. The mnemonic or memory-assisting aspect lies in the similarity of the word and the code so that each brings the other to mind. Geologists tend to use codes of this type informally and inconsistently, although some oil companies, surveys and other organizations work to a standard set of abbreviations, in some cases with computers in mind. For some computer and cartographic applications it is helpful to have a fixed maximum code length, perhaps of four characters to fit one computer word. The mnemonic code may be acceptable as output as well as input, thus avoiding the need for translation in the computer.

Where input is in free format, with words written in full, it may be desirable to reduce the words, for internal storage and processing, to a fixed maximum length by automatic abbreviation, while retaining the uniqueness of each term. The best method to adopt depends, as always, on circumstances. A simple method is to retain only the first *n* letters of each word, but the resulting contraction is likely to be ambiguous. For instance, the same four-letter code would be given by GRANite and GRANodiorite. The Franklin coding method and variations of it (see Brisbin and Ediger, 1967) defines unambiguous and repeatable procedures for creating abbreviations of any fixed length which have a good chance of being unique. The procedures make use of the fact that certain letters and letter combinations are more likely than others to give a pattern uniquely characteristic of the word. The characters which do not have this property are dropped first. The Franklin encoding system specifies a complete set of rules for dropping various letters in turn until the fixed maximum length is reached. A simpler method, which may illustrate the point, is to omit all vowels, unless they are the first letter of the word, and to drop one of adjacent pairs of characters. Th rslt my b qut rcgnsbl. In some applications it may be necessary to sort the codes into an alphabetical order similar to that of the original words, in which case the first few letters at least should be retained unchanged. An example of a simple sortable code for geologists' names is to retain the first three letters of the surname followed by the first initial. The codes are likely to be unique in a group of about fifty English names, and may be preferable to initials as a short identifier, since on sorting they come close to the correct alphabetical order. This procedure occasionally produces an offensive code, and does not work well with Scottish or Japanese names. As an example of the versatility of the coding procedure, the phonetic code may also be mentioned, which is designed to overcome the results of bad or inconsistent spelling. Only sounds which are reasonably constant are represented, and similar sounds given the same code letter, so that Smith, Psmyth, Snaith and Smythe all become SMTH. Where, as in this case, there is considerable overlap of codes additional information, such as day of birth or mother's maiden name can be added.

The generated code or hash code has a special purpose in some data base work. It is a number, calculated on a key field in a record, which determines where that record is stored. A calculation, based on what is sometimes known as a 'hashing algorithm', gives a more or less random number within the range of storage addresses of the records. The best hashing algorithm may depend on the form of the key field and on the characteristics of the computer, but a simple concrete example might be as follows: About 1000 records are to be stored, containing sample information with the sample identifier of two letters and three numbers acting as the key field. A possible hashing algorithm would be to add the binary representation of the two letters, multiply the result by the three-digit number, express the product as a 32-bit integer, retain only bits 2 to 11 and regard this in integer form as the required record number. On searching for the record, the appropriate address can be found directly from the key field by performing the same calculation and thus obtaining the same record number. The procedure can give rise to duplication, in which two different keys give the same address. This can be resolved either by using the next available free space, or by going to a specially designated overflow area if the indicated location is occupied. The full key field is stored, and on searching the keyfields are compared. If they do not match, the overflow procedure is followed until a match is obtained.

For internal storage, it may be desirable to code information in a meaningful way, using what is known as a semantic code, so that items can be arranged in an appropriate sequence or brought into similar groups. The drawbacks to using such codes for recording data were mentioned above, but they are less important in storage and processing formats than they are in input formats, as the lack of mnemonic value and the inhospitable array are no longer so significant. A table lookup procedure can convert from the input to the storage format and provides a means of detecting undefined input codes. Several French groups, in particular, have studied semantic coding in geological applications. One approach is to assign a meaning to each bit in the computer code. A computer word of 32 bits, for example, could be used as the basis for a set of codes for igneous rock types. Each bit in turn would refer to a separate question which could be answered yes or no, being set to 1 if the answer is yes, and otherwise set to 0. The questions might be: 1. Is the grain-size known? (If so set bit 1 to 1). 2. Is the rock coarse-grained? 3. Is it medium-grained? 4. Is it fine-grained? 5. Does it contain more than 50% silica? 6. Are plagioclase feldspars more abundant than orthoclase? and so on. Once the appropriate look-up tables are available, rock names such as granite or basalt could be translated into bit strings which make explicit much of the implicit information contained in the name. Data sets containing semantic codes of rock types can then be searched by computer for combinations of characteristics which are not apparent in the input codes. For instance, it might be possible to investigate whether medium-grained igneous rocks with over 50% silica are commonly associated with igneous rocks in which orthoclase feldspar is more abundant than plagioclase. The ability to search the data for bit-patterns which represent a combination of characteristics allows considerable flexibility in retrieval. A rather similar technique has been tried by Gover *et al.* (1971) in which records were stored as long bit-strings to represent a wide range of concepts in the description of lithological units. The geologists concerned were able to use a rather limited vocabulary of about 200 words in describing rock units, and each bit position in strings of 200 bits was associated with one word.

Semantic codes are not restricted to bit level, and successive bytes in a computer word can be used to carry information, coded from 0 to 256, about various facets of a classification, such as the age range, environment and biological grouping of a fossil. Searches can then be performed on any of these characteristics and not just on the fossil name. Complex, special-purpose software is the price that must be paid for the analytical power that such codes can give. Semantic codes become out of date if the classification to which they refer is altered, and the considerable difficulty of updating semantic codes in a large archival data file suggests that at this stage it may be wise to restrict their use in storage format to small, active data files.

Codes are most often used with non-quantitative data, but may also be required in quantitative fields to indicate missing data items, and items below, or above, the detection limit of the measuring apparatus. If the data are to be processed by FORTRAN programs, a numeric value which does not occur as a real measurement in that field can be used as a code, such as $-1 \cdot 0$ to indicate missing data in a field with measurements of silica percentage. Some statistical processing systems accept absent data in the processing programs, others require that records with absent data should be removed, or missing values replaced with estimated values, before calculation. Values below the detection limit of the apparatus are frequently encountered in geochemical data. Statistical methods to handle such data as a truncated distribution are available, in which a code number for 'less than detection limit' is necessary, and the value $0 \cdot 0$ may be used as a numeric code. With some forms of analysis, it may be adequate to regard the true value as zero, but logarithmic transformation would replace zero with minus infinity, and that is seldom desirable.

In summary, the set of symbols and strings of symbols used to represent non-quantitative data are referred to as codes. There are many types of coding system, designed to save time in writing or punching, save space on a document or map, to assist memorization of the codes, to make them more easily readable, to minimize errors or make the errors more easily detectable, or to carry information implying a classification or relationship. A suitable set of coding procedures can simplify the recording and the handling of data. It is possible to translate between codes for input, storage, processing, communication and output, but too much ingenuity can lead to an unworkably complicated system, and a multipurpose code can be labor-saving. As data files and data systems become larger, an increasing number of geologists work in areas where coding and format conventions are already specified and can economize in programming and systems design effort, as well as making their data more widely usable, by following existing conventions. Particular attention has to be paid to consistency of coding in key fields which provide the means of cross-reference between files.

2.15. Vocabulary control; the thesaurus

Within a large organization, or in a field where computers are extensively used for communication, vocabulary control is important to ensure that all contributors to and users of a data file use the same terms in a similar manner. For this purpose, lists of approved codes may be circulated and updated at frequent intervals. The meaning attributed to geological terms is of course the most important aspect of consistency and has been a preoccupation of geologists for some time. It is not specifically a computer problem, but it may be desirable to maintain a set of operational definitions, setting out the procedure for measurement and recording of each variable. The aspects of vocabulary control more closely related to computer data files are generally handled by means of a thesaurus. This contains all the approved terms, arranged in alphabetical order for ease of manual searching. It also contains synonyms (FOR olivine bekenkinite use analcite jacuparangite), broader terms (BT), narrower terms (NT) and related terms (RT). Thus, under 'arenite' might be noted BT clastic sediment, sedimentary rock; RT sandstone; NT calc-arenite. This assists the searcher to expand his enquiry if he is unable to find all the references he seeks under one term, or to focus on more specific areas through the narrower terms if too many irrelevant references are retrieved. An interesting example of a thesaurus for geoscience documentation is 'Geosaurus' (Lea et al., 1973).

If a data file contains a large amount of uncoded text, as for example in abstracts from papers and reports, or descriptions of outcrops or boreholes, a set of terms, such as brachiopod, andesite or Carboniferous, may be designated as keywords because of their importance in indicating the subject matter. Stop words may also be designated, such as the, and, of, which do not indicate the subject matter and are therefore to be ignored in a computer search. Several geological journals require the author to submit a list of keywords or descriptors after the abstract of a paper, to indicate its scope and subject matter. A similar procedure can be adopted in indexing maps or borehole descriptions. Descriptor lists may help an indexer to classify an article more precisely than he could from the title or abstract. In some indexing systems, the descriptor list is stored directly on the computer for information retrieval. In other systems, the title or the abstract is stored. If the journal does not require a descriptor list, therefore, the author may be able to reach a wider group of readers by ensuring that specific and relevant terms appear in the titles of his papers.

2.16. Data management; the data library, data base and computable model

Aspects of coding are important in recording individual fields; format and data structure are important for sets of complete records. In large and complex data collections, the organization of records into files, sub-files and data banks must be considered. Small data collections are usually tied closely to processing programs, but with larger collections data management is a task in its own right and a data management system is required to handle such tasks as opening new files; storing, sorting and merging data; deleting; inserting, modifying and updating records and items; searching and selective retrieval of records and files. A 'data administrator' may have to be designated for large collections of data to ensure that these tasks are carried out methodically, that the files are adequately designed and documented, and stored efficiently on the computer. Precautions must be taken to ensure that two or more backup copies of the data are available in case the primary record is accidentally erased. Data management programs provide an interface to data analysis and display programs, but where a data management system is used, it is its conventions which determine the mode of recording data, rather than the data analysis system. Just as separation of the concepts of hardware and programs brought greater flexibility and power to computing, separation of data management from other aspects of processing brings greater flexibility to data handling and enhances the geologist's ability to deal with large and complex sets of data. Data management is an active and specialized branch of systems analysis, and is considered again in the chapter on communication. At this stage, however, it is appropriate to deal in greater depth with the organization, structure and content of large collections of complex data (see Deen, 1977).

The term 'data bank' has been used in many ways, to mean many things, and perhaps the term is now most useful in a non-specific sense to mean an organized collection of data, probably accessed by computer. The simplest form of data set for data management purposes is a file collected for analysis by a single program, and which is of no further value after the analysis is complete. Considerations of data banking do not then apply. The simplest form of data bank reflects a need to store data in computer-readable form for possible later analysis. A collection of independent, though not necessarily unrelated data sets or data files, collected within various projects with differing objectives and procedures, constitutes a data library. The reason for storing the data might be that, even if they had already undergone extensive computer analysis, the raw data are the evidence with which other workers can independently verify the conclusions by re-analysis or check the data collection procedures by re-examining a small sample. The development of new methods of analysis, new data or new theories might also lead to new examination and analysis of the raw data. A computer data file can thus be a real contribution to scientific knowledge, in a similar way to a descriptive scientific paper. As with the latter, the expected usage of the raw data might be very small, and it would not be surprising if many data sets were never re-examined. The useful life of the data, on the other hand, could be very long, just as a paper could be of considerable interest which described the geology of an area that was last studied 100 years ago. Craig (1969) suggested that 'publishing' some data sets, in the sense of making them available to the public, could be achieved more cheaply, and probably more usefully, through a computer data library than through conventional publication. The contributor of data to a data library, like the author of a paper, has the responsibility of making appropriate decisions about data content and format, while the 'editor', in charge of accepting material for the data library, has the task, with the help of referees if need be, of ensuring that the contribution is relevant, is not offensive, meets the house style, and is an original and worth-while contribution to scientific knowledge. For the information to be useful to others, some descriptive material is clearly necessary. In the case of contributions to a data library, it is likely that a file will be linked to, and will supplement a scientific paper of conventional type explaining the objectives, sampling methods, methods of data analysis and conclusions. However, it is also useful to store some descriptive material with the computer data set, so that its content and arrangement are clear to other users. The approach to this depends on whether the data set is an array of similar records, in which case a description of the record type may suffice, or whether each record contains different information, in which case an account of the record structure and a summary of the information content is required. Examples of the first approach are provided by the ROKDOC (Loudon, 1974) and G–EXEC (Jeffery and Gill, 1976) data descriptions, in which it is assumed that the data are recorded as two-dimensional arrays, with each record containing similar information. An example of the second approach is provided by Cutbill et al. (1971).

ROKDOC, which was a system designed for small geological data libraries held on punched cards, provides a typical example of a data description. The data description (or DD) is a set of data records which precede the data proper, and record such information as the name and address of the contributor of the data, a title for the data set and a brief description of its contents, indicating for example, the geographical area and stratigraphical formations from which the material was collected, the topic under consideration, and the aspects of the geology about which information is recorded. The format of the data cards and the number of variables is indicated. For each variable in turn, a short mnemonic code for the variable name is recorded, followed by the name in full, with brief comments on the variable, such as the units in which it was measured. Where variables are recorded in encoded form, the codes are listed in the data description, and a one-line explanation is given of each. The G–EXEC data handling system, which was developed with larger data sets in mind, adopts a more flexible approach by storing the data description with the data in the computer, the description being updated automatically when the data are updated. Codes in G–EXEC are held in separate 'dictionary' files, linked to the appropriate data sets through the data description, thus making it possible for several data files to be linked to one dictionary. Special codes, indicating missing data or end of data are stored also in the DD.

The computer can obtain information from the data description for labelling variables in the output, reading the data correctly, and recognizing codes. For the user, the data description ensures that enough descriptive information is available to him to decide whether the data are relevant to his purpose, and even for the contributor can be a useful reminder of what was recorded and how it is arranged. As with any form of publication, a prime requirement is availability of the information. If data libraries are to be successful for data exchange on a large scale, they must share the existing well developed mechanisms for communication developed for bibliographical information. Each data set must therefore have the equivalent of a bibliographical reference, namely: date of publication—date of acceptance by the data library; one or more authors—the contributors of the data to the library; a title; the serial name, volume and number—the name of the data library or referral center and the file number. In addition, an abstract or a set of keywords describing the content might be required for information retrieval. This bibliographical information for a data set has been called a 'label', and in principle can be used in the same way as a reference to a book or journal, being referred to in the reference section of a book or paper, and included in library current-awareness and information retrieval systems. The function of a data library is to archive data in computer-readable form and make copies of data sets available on request. A referral center has the somewhat different function of maintaining and publishing lists of data sets, but referring enquirers to the individual or organization holding the data (see Oppenheimer et al., 1976). Data which are recorded following the conventions of an integrated system can access the various programs in the system directly, for data editing, management, processing or display, and although in principle a data library could accept input in any reasonable format selected by the contributor, in practice it may be desirable to define the input formats rather closely, in order to make it easier for the user of the data to find appropriate programs for analysis.

Some flexibility of data structure is achieved in G–EXEC as in some other systems through the use of so-called 'index' files. Their function is to establish an explicit link between one or more records in one file, and one or more records within one or more other files. This makes it possible to retain the simplicity of the two-dimensional array for recording data, with the flexibility of a hierarchical structure linking records between files or subfiles (see Fig. 27). By recording the relationships in a file separate from the data files, additional freedom is gained to alter or extend the linkages independently of the data. When the data are first recorded for the computer, the links may be implied either through the arrangement of the data, for example, the location and name of a borehole might precede the description of the formations which it penetrated, or through a cross-reference key field, such as the borehole identifier, being stored with each formation description. A convenient input structure can thus be mapped into a convenient storage and processing structure.

This rather simple example illustrates the ability of the computer to handle complex relationships, and this ability has been further developed with efficient means of linking individual data items. This is part of one major recent development in computer science, namely, data base management. As a concept, the data base is at the opposite end of the data bank spectrum from the data library. An actual data bank is likely to lie somewhere between the two extremes as shown in Fig. 28, and to have some of the properties of each. Whereas the data library is concerned with independent contributions, each submitted on its own terms, and requiring little software support other than standard utility programs for tape copying, the data base is concerned with interrelated material contributed according to a predetermined framework designed to make the data base available to many users for many purposes. The data base concept seems to have gained acceptance first in commercial applications where separate departments within an organization created files of information on such aspects of a firm's activities as salaries and wages, personnel records, sales, invoices and receipts and stock. By redesigning the files and integrating them as one data base, new applications became possible, such as using the sales and the personnel files to calculate salesmen's commissions to pass to the salaries file. New possibilities were also created for producing new and more comprehensive management information. The value of these concepts in commercial applications has now been fully demonstrated, extensive data base management software is available from computer manufacturers and

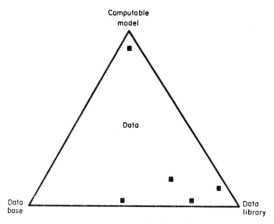

Fig. 28. The data bank triangle. A data bank is likely to have some properties of each of the three ideal end members to a greater or lesser degree, and can thus be shown at an appropriate position within the triangle. Reproduced with permission from Oppenheimer *et al* (1976).

software houses, and a coherent rationale and terminology have developed. The application of the ideas to the environmental sciences is a slow process, but one which is well under way in some large organizations. Since the concepts may be relevant even in quite small projects, they are worth considering in more detail. Some of the ideas can be put into practice, without becoming involved in a comprehensive data base management system.

A data base, in contrast to a data library, contains a single, integrated collection of data, in which relationships between items are explicitly indicated. The pattern, structure and relationships of the data are defined before they are assembled in the data base, and redundancy or repetition of data can therefore be controlled. If a data item is used in several contexts, it may nevertheless be stored only once, and linked in through cross-references as required. When a data item is updated or amended, therefore, only one entry need be altered, and all cross-references automatically refer to the most recent material. Because data are held in only one set of master records, consistency between files is easier to maintain. If four separate files in a data library, for example, referred to the paleontology, petrology, geochemistry and geotechnical properties of material at a locality believed to be Carboniferous in age, then each file would be in error if the rocks were subsequently found to be Permian. In a data base, only one entry need be changed. A data base is designed for ease of change and updating, and if it is to develop systematically, it must be under the active, day to day control of one organization and one group. A data library, with a scattered community of contributors, cannot be updated in this way. Indeed, like the

H

scientific literature, it is fundamental that information published at a particular time should be kept unchanged. If a published scientific paper is found to be in error, or views change, the paper is not altered. Instead a new paper is published, putting forward the revised view. The data library, as a mechanism for data storage and exchange, plays little part in determining what data are collected. The data base, on the other hand, requires tight control over both the content and quality of data submitted, and the data are recorded to meet a requirement set down in the data base specification. Organizations such as geological surveys, government departments, oil companies, mining houses or museums are concerned with maintaining the non-computer equivalents of geological data bases, such as series of geological maps, well records systems, reports or museum catalogs, and they are most likely to be involved in establishing and maintaining a computer data base. Some data base concepts are applicable whether or not a computer is used, such as controlling duplication of data, maintaining consistent identifiers and cross-references, and attempting to store the data in the most useful and readily available form, with adequate backup copies for security, and indexes for efficient access. Research undertaken by individuals or small groups in universities or elsewhere is more likely to result in data files that could be contributed to a data library.

A number of steps in the transition from data library to data base can be visualized. First, the files in the data library could be treated as working files, with provision in the software for updating and modifying records. At least three levels of access might have to be defined, since different groups or individuals might have the authority to read, update or modify the file. Membership of the group would have to be known exactly and would probably not be the same for each file. The software would have to include provision for prevention of accidental or deliberate access to the file by an unauthorized user. A central group would be required to maintain a thesaurus and lists of codes, particularly where these overlapped several files. As a further step in the transition to a data base, a data bank could include index files for cross-referencing between files, and again the responsibilities for these would have to be decided, since different groups would be concerned with different patterns of file linkage, and some of the index files might be established some time after the original data files. The validity of the cross-referencing would then have to be considered in detail for all relevant variables, to ensure that appropriate data from several files can be matched and correctly analyzed. For example, it might be necessary to ensure that geographical coordinates were measured in the same way in linked files, that the same sample reference numbers were used throughout, that geochemical analyses referred to comparable samples to those on which the petrography or physical properties were investigated, and that porosity measurements for cores were adjusted to the same depth scale as the geophysical logs. Some controlled redundancy might be necessary for efficiency, such as establishing a rapid access file repeating some of the frequently used data drawn from various files. Inverted files, in which are listed all occurrences in the data base of, say stratigraphic names, arranged in sorted order, might be required to provide efficient access on certain defined attributes frequently used for searching. Reduced searching and retrieval time is gained at the cost of increased up-dating and editing time. Dictionary files would be required to record the content of the files, with special precautions to ensure that all occurrences of an item were amended. After these facilities are introduced, the function and usage of the data bank would be close to that of a data base, although the design is that of a modified data library.

The Codasyl Committee, which is concerned with aspects of standardization in computer languages and related topics, set up a Data Base Task Group, which prepared a major report in 1971 (Codasyl, 1971), considering the data base in its own right, rather than as a modification of existing file management methods. It brought together ideas from a number of sources, and has considerably influenced subsequent developments. Data are not considered in terms of files, but as individual data items, which are defined as the smallest units of data in a data base. A data item can be referred to by name, such as 'well identification number', and has defined properties such as field length and data type. One or more related data items constitute a data record, and a record type is a named description of a collection of data items which occur repeatedly in the data base. Thus, 'well reference data' might be the name of a record type containing the well identification number, the well name, driller, location, ground and KB elevations and total depth. A set is a named relationship between several record types of which one is designated the 'owner' and the remainder the 'member' record types. Conventionally, the set can be illustrated by several boxes representing record types, with arrows representing the relationships between them, pointing from the owner to the member record types (see Fig. 29). The complete description of all data items, record types, sets, etc. is known as the schema (pronounced skeema) and constitutes the data base description. It is an intrinsic feature of the data base approach that a single data base can be viewed in different ways for different applications. This is achieved by the use of the 'subschema', which is the description of that part of the data base which is required by and is visible to a particular application, and names the record types, set types, etc., that are involved. Many subschemas, but only one schema, may refer to one data base. Entry to the data base by application programs can be achieved efficiently through specified data items located by a hashing algorithm. The calculated entry points which can be accessed in this way are shown on the data base diagram (Fig. 29) by asterisks.

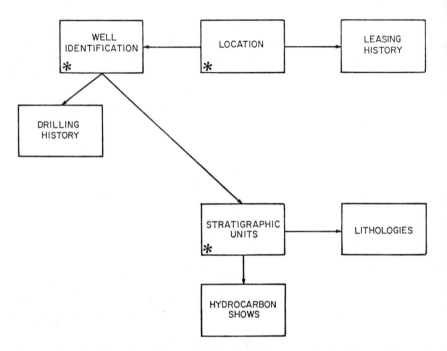

Fig. 29. Record types and sets in a small data base. Entry to the data base can be made through keys in the records marked with an asterisk.

For instance, well identification number and sample number might give suitable entry points to a data base, percentage of sodium would not.

A data description language provides the means of defining the schema and subschemas, and ideally is independent of the language for accessing the data base. A data manipulation language contains instructions for the operations of storing, finding, deleting, or modifying items in the base. In some implementations of the Codasyl proposals, these functions are seen as extensions to the COBOL programming language, reflecting the greater importance of data bases in business as opposed to scientific programming. Similar instructions can be incorporated within FORTRAN or PL–1, by implementing them as subroutines, invoked as required in the program by CALL statements. The data manipulation language thus becomes an extension to several host languages, and in principle the same data base may be accessed from application programs in more than one language, while the data description language is separate, independent and self-contained.

A number of data base management systems (DBMS) are available from the computer manufacturers and the so-called software houses, firms specializing in the preparation and marketing of computer programs. Implementation and maintenance of a full DBMS is a large overhead which is justified only for large data bases of lasting importance, although many of the concepts are relevant to smaller tasks, even where no computer is used. Geologists may, however, be able to utilize a system implemented and supported for other purposes. Many of the DBMS follow the Codasyl proposals, others follow different definitions and approaches, since uniform standards cannot be expected at this early stage. One divergent line of development (Codd, 1970) views the data as a set of 'relations', or files in simple but strictly defined tabular formats, linked through keyfields. Relational algebra defines operations for retrieving and managing data in the so-called relational data base. The applications are normally considered within a commercial or business context, but the complexity of geological relationships suggests that some of the ideas could also be important in geology. Within the conventional data base, the relationship of set membership can be expressed naturally, such as the fact that lithological units are members of a set of such entities within a single borehole. A geological map, however, illustrates many relationships in space and time. The relationships between two formations shown on a geological map might include the following: is adjacent to; is underlain by; follows unconformably on; is cut by; is near to; is penetrated by; is approximately in the position of; passes by transition to; is separated by a fault from; is contained within; interfingers with; has a diachronous boundary with; etc. Many other complex relationships exist between a set of boreholes and the formations they penetrate, or between geophysical reflectors and geological horizons, or between chronostratigraphic, biostratigraphic and lithostratigraphic units, and no satisfactory notational system has been devised either for the computer or for conventional communication. This may prove to be a fruitful area of future research.

The data base, as opposed to the data library, has been considerably more successful in the business world than in geology. There are many reasons for this, and the tendency of geologists to work as individuals rather than as large organized groups may be an important one. The more fundamental reasons, however, may lie in the subject matter and in the nature of the data. A data base for a commercial organization might contain information on personnel records, salaries and wages, equipment inventories, parts numbers, stock levels and production figures. Such information refers to data items that can be precisely identified, and there are clearly defined relationships between them. Any doubts about who John Smith is, or what his salary and tax deductions should be, would be speedily resolved. Geologists are concerned with comparatively fuzzy concepts, like the 'level of the top of the Carboniferous', or 'intruded granite' of 'marine facies' which are difficult to define unambiguously. Geology is concerned with

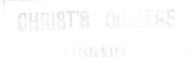

processes that took place in the distant past, and with the results of these processes, which are unlikely to be fully accessible to observation. Even the accessible geological evidence is so vast compared with the resources available to study it, that only a very small sample can ever be examined. Collection of data for a completely general range of objectives is thus not possible in geology, and each data set must to some extent reflect the reasons for collecting the data. Therefore, full integration at the data base level is not possible, since the links between the data sets exist only in the background of geological hypothesis and theory. A geologist may tie his own observations together in terms of, say, facies distribution and surface analysis, that is, with entities which he knows only by deduction and not directly from the data. The historical geology of a region may be the level at which individual sets of data and the conclusions of several workers are integrated into one coherent picture. In the longer run, as suggested in the chapter on explanation, the background information which controls data collection and the relationships between data sets may be represented as a computable model, and data collection and storage can then be seen as part of an integrated model-building activity. Meantime, the model is largely outside the computer, on paper or as assumptions in the geologist's mind, and the data library, rather than the data base, is well suited to this situation. With techniques already in use, a data bank may include or be linked to algorithms for interpolation of points on a geological boundary by splining, or for silicate re-calculation from geochemical analysis, or interpolation between boreholes to predict the value of points on a surface. Thus, data are generated rather than retrieved, using known data points on the one hand, and the geologist's view of the best predictive tool on the other. The data bank thus has some of the characteristics of a computable model, and the concept of the spectrum from data library to data base can be extended as in Fig. 28, so that a data bank can be plotted as a point which indicates the relative importance of the three end members.

3. INTERPRETATION

3.1. Objectives in interpreting data

In the model of the system of geological investigation, the first subsystem, of observation and data collection and recording, was seen as an interaction between the geologist and the real world, and the channel through which primary observational and factual information passes to the second subsystem—data interpretation. The interpretation of data is the subject of this chapter, and is seen as encompassing the various ways in which the geologist can manipulate and transform his data to help him to appreciate and understand their properties. This is a somewhat different activity to explaining his observations, which is a process of arriving, by analogy and intuitive reasoning, at an account of the origin and development of the observed geological features that can be reconciled with the existing framework of geological knowledge. Explanation calls for a different kind of mental process (and computer procedures) to that required in interpretative data analysis, and is therefore considered later as a separate subsystem. Data analysis, as considered here, is concerned with the extraction of as much useful information as possible from a set of data. It is a closed system, in that no information is added. It may, however, make information available which, although inherent in the original data set, was not apparent to the geologist.

The computer is, in practice, essential for many techniques of geological data analysis, in such fields as multivariate statistics or time series analysis, and can also be useful in more traditional areas like the preparation of cross-sections and geological maps. It is in the ability to obtain appropriate data, to select the best methods for their analysis and manipulation, to display the results to best effect and to draw appropriate conclusions in full awareness of the limitations of the data and the methods, that the interpretative skill of the mathematical geologist lies. Computer methods put a large battery of techniques at his disposal, in which the detailed mathematics are already codified as computer programs. The aim of this section is to show that there is a straightforward logical pattern to what is, at first sight, a bewildering variety of techniques, and to show the part that they can play in a geological investigation. Three relevant concepts, which were introduced earlier, are: frame of reference; redundancy; and transformation and translation.

3.2. Frame of reference

A frame of reference is the context in which information is interpreted. The human mind considers information that it receives against a background of expectations and past experience, comparing new data with established reference points. Thus an examination paper submitted by a high-school or university student is judged against the background of his past work and the work of his fellow students. When the same individual has acquired some years of experience as a practising geologist, and submits a paper to a scientific journal, the editor considers it in a different frame of reference, on the basis of the quality and style of papers normally published in that journal. In less important examinations, the student may be compared with those in his own group. In awarding a University degree, a wider frame of reference is adopted, and the comparison may be national standards over a period of years. A changed frame of reference can be informative in making relationships clearer. The speed at which a car normally travels on the highway is vividly appreciated if it runs off the road into a rough field or scrubland normally seen at walking pace. The moon seems larger than usual when it is low on the horizon and seen against a familiar background of trees and roofs of houses. The detailed geomorphology of an area as studied on the ground may take on new meaning when viewed from an aircraft, and pattern may be revealed in structural measurements for the first time when they are seen plotted together on a map. One of the strengths of statistical methods in regional geology is the ability to calculate measures of local variation within a local frame of reference, while at the same time considering variation between subareas within a regional frame of reference. Another strength of statistical analysis is the ability to alter the frame of reference in terms of geological properties. Using traditional methods, the geologist is accustomed to thinking of the distribution of rock types in present-day space, then altering his frame of reference to consider the distribution of depositional environments and facies at periods in the geological past. With computer methods, the geologist has greater freedom to explore the relationships among many variables and spatial patterns in the original data, or in derived variables which may bear a closer relationship to the geological processes which created the rocks. He can view the available information in the frames of reference that are most relevant to his purposes.

3.3. Redundancy

Redundancy in a system is concerned with duplication of a function or a component. A car may have a spare tyre and a second, separate, emergency brake. A computer may have two line printers where one would do. The redundant components are not essential to the operation of the system under normal circumstances, but may be deliberately included as a standby in case the original component should fail. Deliberate redundancy can be desirable if a component is particularly likely to fail (the spare tyre) or where the consequences of failure would be particularly serious (the emergency brake). Redundant information is information that could be recreated from other available information. An application form which requires the date and the candidate's date of birth need not include his age. It may nevertheless be requested as deliberate redundancy, since a quick check that the three were in agreement could reveal an accidental error. The introduction of redundant information is an important aspect of controlling errors, and examples mentioned previously include the parity bit, the check-sum and the error-checking code.

Redundant information is also likely to arise in the collection of data, not through any deliberate intention on the part of the geologist, but because he is measuring correlated variables which show regularity in their distribution in space and time. The position of a straight line in space can be described by two points, and the position of additional points on the line would be redundant information. Similarly, the regularity of many geological surfaces means that they can be described by a limited number of points. An example of this is provided by the process of digitizing geological boundaries shown on a field map or reflector horizons shown on a seismic section. From a comparatively small number of carefully selected points, it is possible, with appropriate methods of interpolation, to recreate the original lines with a degree of accuracy. The rest of the line was thus theoretically redundant. Where a large number of geological variables are measured, they reflect, to varying degrees, the influence of underlying controls in the geological environment, and thus tend to be correlated with one another. The values of some variables can therefore be predicted from a knowledge of the behavior of others, and part of the information is redundant. Removal or reduction of the random element reduces the amount of information that the geologist has to consider. It may also reveal interrelationships and patterns in space more clearly, separated from the haphazard background variation, which is redundant in the context of geological interpretation. The redundant variation is sometimes known as 'random' variation to differentiate it from the regular, predictable pattern, and to emphasize its uncertain, chance occurrence. Some writers prefer to draw an analogy with communications theory, and refer to the unpredictable element as 'noise', as opposed to the pattern or 'signal'.

3.4. Transformation and relationship

The third concept that needs some introductory consideration is that of transformation. Transformation is conveniently defined in terms of two sets, A and B. A transformation, sometimes known as a function, correspondence, operator, or mapping, from A to B, associates each element of the set A uniquely with an element of set B. The logarithmic transformation, for instance, would associate each member of the set of real numbers, with the corresponding member of the set of logarithms of these numbers. Other familiar transformations are the trigonometric functions, such as the sine or cosine, the inverse function in which x is associated with $1/x$, the exponential and the square-root function. The motive for transformation in the present context is generally to replace a set of data values by transformed values that are more easily manipulated. The mechanism for performing the transformation on the computer may be a program or subroutine which refers to a correspondence table, like the sine tables of elementary trigonometry where one looks up the value of x to find the value of sin x, or through a subroutine which uses an algorithm to calculate the value, for example, using the series.

$$x - x^3/3! + x^5/5! - x^7/7! + \ldots$$

to calculate the sine of x. The idea of correspondence between sets has general significance in computer applications, and transformation from a set of mineral names to their semantic codes, or from a sample to its geochemical analysis or petrographic description are clearly important operations in interpreting geological data.

The idea of relationship, which was mentioned as an aspect of data structures, is close to that of transformation. A relation, in the framework of set theory, is concerned with the existence or otherwise of some type of bond between pairs of objects. The relation could be, for example: 'is the semantic code for'; 'is the sine of'; 'is the silica content of'. The transformation between two sets may thus be an expression of the relation between members of the two sets. A particular kind of relation, known as an equivalence relation, can be illustrated by examples from geometry. Congruence is a type of equivalence between geometrical figures, expressing the fact that two figures, such as the triangles shown in Fig. 30, are identical except for their position in space. Similarity is a weaker form of equivalence in which figures, such as the triangles of Fig. 31, are of the same shape, but not necessarily of the same size. An equivalence relation is a relation which satisfies the following

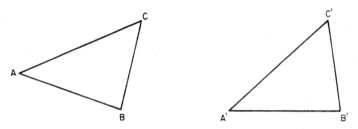

Fig. 30. Two congruent triangles, illustrating an equivalence relation in geometry.

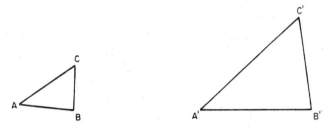

Fig. 31. Two similar triangles, showing a weaker form of equivalence relation than that in Fig. 30.

conditions: an element is equivalent to itself; if a is equivalent to b, then b is equivalent to a; if a is equivalent to b, and b is equivalent to c, then a is equivalent to c. To give an example, if Jack is the brother of Bob, and Bob is the brother of Ted, then Jack is the brother of Ted, and Ted is the brother of Jack. Nevertheless, since Jack is not the brother of Jack, 'is the brother of' is not an equivalence relation. Again, 'is the father of' is not an equivalence relation. If Joe is the father of Ted, it does not follow that Ted is the father of Joe. However, 'lives in the same house as' or 'is in the same family as' would be equivalence relations on the set containing Joe, Jack, Bob and Ted.

3.5. Distance and metric; geometric and algebraic operations

An important transformation which connects two members of the same set with the set of real numbers is that of distance. Distance, in everyday usage, refers to the relative geographical position of two points in three-dimensional space. In a more abstract mathematical sense, the meaning is extended to refer to a transformation connecting any two members of a set with the set of real numbers, with the following properties: the distance from a to b is the same as that from b to a; the distance from a to b is zero if, and only if, $a = b$; and the distance from a to b, added to that from b to c, is equal to or greater than that from a to c. In the simple geometric case, this reflects the idea that the distance between two points A and B is the same regardless of whether it is measured from A to B or from B to A; that a point coincides with itself; and that the sum of the lengths of two sides of a triangle ABC, that is, AB+BC, is never less than the length of the third side AC. AB+BC = AC if and only if B lies on the line AC. Distance as defined above is, however, a much more general concept than geographic distance, and one could, for example, define the distance between two chemical analyses or two fossil species. Where the formula for calculating the distance between each pair of elements in a set is defined, it is known as a metric, and the set is said to occupy a metric space. It is possible to define more than one metric for one set, and the set is not, therefore, limited to a single metric space.

A basic feature of much geological data analysis is that the methods are expressed algebraically for implementation on the computer, but in most cases they can be visualized geometrically. The ability to move easily between a geometric and an algebraic formulation is the key to understanding many of the expositions of data analysis methods in the geological literature and is emphasized throughout this chapter. The geologist is accustomed to visualizing points, lines and surfaces in two- or three-dimensional space, and is unlikely to have difficulty in considering space in terms of three axes at right angles, sometimes referred to as orthogonal or rectangular axes, oriented, say, east, north and vertically upwards (see Fig. 51). In each of these directions, a zero point, datum or origin is defined, such as the bottom left-hand corner of the map, or the intersection of two lines of latitude and longitude, and sea level. Distances from the origin along the three axes specify the position of any point in three-dimensional space, and a reference grid may be based on lines at fixed intervals, say 1000 meters apart, parallel to the axes. This is precisely the framework of three-dimensional geometry, and the algebraic representation is as easy to understand as the geometric.

The coordinate axes can represent any quantitative variable, and, because the axes are at right angles, it is not essential that the same units of measurement should be used in each direction. The percentages of silicon and calcium, and parts per million of thorium could be represented by three axes, and a point in that three-dimensional space would indicate the chemical composition of a rock for these three elements. Being accustomed to living in a three-dimensional world, it is not easy to visualize space of four or more dimensions. But there is no need to visualize it. Relationships are expressed algebraically for computation, and there is no difficulty in expanding equations to any number of terms that is required. Where the number of variables is n, the data are said to be represented in n-dimensional space. Since the principles and procedures are not affected by the number of dimensions, the three-dimensional geometrical picture remains a useful illustration of the process. The distance function, in its full generality, makes it possible to

relate the various axes within one metric space, and to analyze the properties
of several variables together.

In geographical space, the study of three-dimensional coordinate geometry
has a long history, and most of the techniques required in geological applications were established by the time of Euclid. The calculation of the distance
between points with coordinates (x_1, y_1) and (x_2, y_2) follows directly from
the theorem of Pythagoras. The square on the hypoteneuse of a right-angled
triangle is equal to the sum of the squares on the two opposite sides. Hence the
distance between the two points is the square root of $(x_1 - x_2)^2 + (y_1 - y_2)^2$.
In three dimensions, the vertical coordinate, z, must also be included. The
distance between (x_1, y_1, z_1) and (x_2, y_2, z_2) is equal to

$$\sqrt{[(x_1 - x_2)^2 + (y_1 - y_2)^2 + (z_2 - z_2)^2]}.$$

An additional term is added for each higher dimension. This metric defines
what is known as Euclidean distance, and the corresponding metric space
is known as Euclidean space.

An example of data analysis involving calculation of geographical distance
might be provided by a six-day field trip to a number of fossil localities,
(see Fig. 32). Each locality is identified by a number, and its location

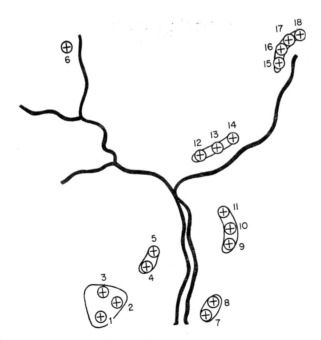

Fig. 32. A map showing the position of eighteen outcrops that might be visited during a
field trip to collect fossils.

recorded in terms of x and y coordinates measured from a map. From this data set, a matrix can be calculated, as shown in Fig. 33, of the distances between each pair of localities. One can determine from the matrix which localities are nearest to each other, and which groups of localities are in the same neighborhood. One method of presenting this information is in the form of a dendrogram, as in Fig. 34. The dendrogram shows distance between items on its vertical scale, and each item (locality in this example) is shown at a separate position along the horizontal axis, arranged to bring nearby points together. The distance between each locality and the nearest neighboring locality is shown by linking them with a cross-bar at the appropriate point on the distance axis. That pair of points is then linked to their next nearest neighboring locality at the appropriate distance value, and so on. If, in the example, it was decided that the field trip should visit a group of adjacent localities on each of the six days, then the six main groups which are apparent on the dendrogram would suggest a means of assigning the localities on each day. The single locality which does not fall into any of the six classes, might be left aside as being too difficult to reach.

It is comparatively easy to visualize distance in geographical space, but the procedure is as general as the definition of distance, and can be applied in an analogous manner, where the metric space can no longer be visualized. The concept of Euclidean distance can be applied to any pair of items on which the same set of quantitative measurements have been made. It is calculated by taking the square root of the sums of squares of the differences of values of successive variables. Table 1 shows the calculation of Euclidean distance between two samples on the basis of geochemical analyses.

TABLE 1.

Element	Si	Al	Fe	Ca	Na	K
Sample 1	30	8	5	4	2	2
Sample 2	24	7	7	5	2	2
Difference	6	1	2	1	0	0
Difference squared	36	1	4	1	0	0

The sum of squares of differences $= 42$

Hence, distance between samples $= \sqrt{(42)} = 6 \cdot 5$

A fundamental objection can be raised to this procedure. In geographical space, position can be measured in kilometers east and north of the origin, and distance can be measured in the same units. But to combine percentage

	1	2	3	4	5	6	7	8	9	10	11	12	13	14	15	16	17	18
1	0·0																	
2	0·4	0·0																
3	0·4	0·3	0·0															
4	1·2	0·8	0·9	0·0														
5	1·5	1·1	1·3	0·3	0·0													
6	4·7	4·6	4·3	4·1	3·9	0·0												
7	1·9	1·6	1·9	1·4	1·5	5·3	0·0											
8	2·0	1·7	2·0	1·3	1·4	5·1	0·3	0·0										
9	2·6	2·3	2·4	1·5	1·4	4·5	1·3	1·0	0·0									
10	2·8	2·4	2·5	1·6	1·4	4·3	1·6	1·3	0·3	0·0								
11	2·8	2·5	2·6	1·7	1·4	4·0	1·8	1·6	0·6	0·3	0·0							
12	3·3	3·0	3·0	2·2	1·9	3·0	2·8	2·6	1·5	1·4	1·1	0·0						
13	3·6	3·2	3·2	2·4	2·1	3·2	2·9	2·7	1·7	1·4	1·1	0·3	0·0					
14	3·9	3·5	3·5	2·7	2·4	3·4	3·1	2·9	1·8	1·6	1·3	0·7	0·3	0·0				
15	5·4	5·1	5·1	4·3	4·0	3·8	4·6	4·3	3·3	3·0	2·8	2·1	1·8	1·6	0·0			
16	5·6	5·3	5·3	4·5	4·2	3·8	4·8	4·6	3·5	3·2	3·0	2·3	2·0	1·8	0·3	0·0		
17	5·9	5·5	5·5	4·7	4·4	4·0	5·0	4·7	3·7	3·4	3·3	2·5	2·3	2·0	0·4	0·2	0·0	
18	6·0	5·7	5·7	4·9	4·6	4·2	5·2	4·9	3·8	3·6	3·4	2·8	2·5	2·2	0·6	0·4	0·2	0·0
	1	2	3	4	5	6	7	8	9	10	11	12	13	14	15	16	17	18

Fig. 33. A table of distances between every pair of fossil localities shown in Fig. 32.

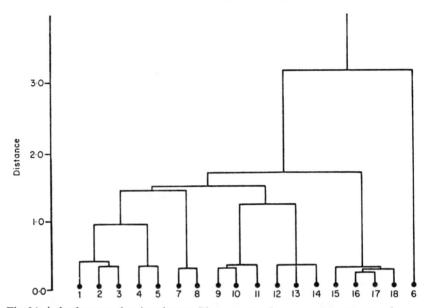

Fig. 34. A dendrogram showing clusters. They are sets of outcrops from Fig. 33 which are close together.

of silicon with percentage of calcium is reminiscent of adding 5 apples to 2 oranges, and trace elements, measured in parts per thousand, cannot be appropriately handled in the same way as the percentage data. In order to use a distance function effectively in these circumstances, a method must be found of ensuring that the distances on the original axes are scaled to reflect the relative importance of the variables. Two simple approaches to this problem would be to consider each variable as equally significant and scale it so that its values all lay between 0 and 1, or to assign a rank or score to each item for each variable, so that the sample with the largest silicon content scored 1 for silicon, the item with the next largest scored 2, and so on. Another, more satisfactory, method, namely, standardization of the data, is described later, but the immediate point is simply to indicate that methods do exist.

Many kinds of distance function have been proposed. For qualitative data, for example, a 'match coefficient' can be used in which 1 is scored for each pair of matching characteristics. This is easiest to explain by an example:

Characteristic	Lithology	Grain size	Structures	Base of bed
Sample 1	SD	C	XBDG	EROS
Sample 2	SD	VC	GRDG	EROS
Match score	1	0	0	1

The total score for the match coefficient between samples 1 and 2 is 2 out of a possible 4, or 50%. The match coefficient is not a distance function as described above, because it is larger rather than smaller when the items are more alike. It is therefore known as a similarity function. It can readily be converted to a distance function by reversing the scoring procedure, but this may not be necessary in practice, since similarity functions can be equally useful. The match coefficient can be made more complex by scoring −1 for a mismatch, and zero when a characteristic was not observed in either or both items. Fractional scores can be used for degrees of similarity, so that the difference between very coarse and coarse-grained sandstone is recognized as being less than that between very coarse and very fine. Mutual dependence between variables can be taken into account by specifying, for instance, that similarity of grain size scores only if lithology is the same. It is also possible to devise distance functions which can handle a mixture of quantitative and qualitative data. The range of possibilities of deriving distance functions is in fact limited largely by the imagination of the geologist. In practical implementations, however, the user may have to be prepared either to accept a standard approach and frame his data set accordingly, or to write his own computer program.

3.6. Weighting of variables

An important concept which arises with distance functions and in other aspects of data analysis, is that of the relative importance of different variables. In the first example of distance analysis described above, a number of geological localities were grouped according to their distance apart, so that a group of localities in the same neighborhood could all be visited on the same day. In mountainous country, the relative altitude of the localities would also be important, and could be incorporated in the data and in the calculation of distance. But the true distance in this case would give misleading results. It is more tiring, and takes longer, to climb 5280 vertical feet than to walk a mile on level ground. Altitude could therefore be exaggerated relative to the horizontal distance, with perhaps 1000 vertical feet being regarded as equivalent to 3 horizontal miles, on the basis of the time taken to cover the ground on foot. This procedure is known as weighting the variables, and is a transformation selectively applied to certain variables to alter their relative magnitude. It can be visualized graphically in the case of the mountainous field trip by thinking of distances on a relief model in which the vertical scale is greatly exaggerated.

There are several ways in which weighting could be handled in the computer. One would be to store the data in weighted form, another to incorporate the weighting function in the distance analysis program. Neither is

I

entirely satisfactory, since each lacks flexibility. Another user might wish to use similar data and the same program to calculate the length of cable required to bring an electricity supply to all these points, and this would require unweighted variables. To give a geological example, a set of geochemical analyses might be grouped to determine which ones suggested the presence of economic ore deposits. A change in the economic values of different ores a few years later might require a new analysis of the data with revised weighting factors. To achieve the required flexibility, the unweighted data must be stored, and the program designed in modular form, so that each individual variable can be assigned a weighting if need be, using a separate program module, before the distances are calculated and the analysis performed.

3.7. Variables in metric space

For a particular data set, the units in which the variables are measured and the method of calculating the distances between points, together define the metric space in which the data are considered. A typical set of geological data might consist of measurements of a number of variables on several items, structured as a matrix. The variables might be percentages of chemical elements in a number of samples; or the dimensions of brachipods in a collection, such a depth, width, length of hinge-line, etc; or thicknesses of formations encountered at a number of wells; or gas/oil ratio, gravity of oil, hydrostatic pressure, thickness of oil column and depth of several oil reservoirs. The variables can be represented by axes in n-dimensional space, and each item defined as a point in that space. If twenty variables had been recorded for 200 items, there would be 200 points in 20-dimensional space. The points might be evenly distributed through the space, but it is more likely that they occur as groups or clusters of more closely spaced points, separated by areas in which fewer points are found, just as stars are not evenly distributed in space, but tend to cluster together as galaxies. The reasons might have geological significance. Geochemical samples might come from more than one source rock, or from areas, some of which had undergone mineralization and others not. Fossil specimens might come from more than one species or subspecies, or from different environments, or from different sex or age groups. Formation thickness might reflect localized but persistent depositional or structural environments. The properties of oil pools might reflect several source rocks or hydrodynamic regimes.

There is a vast amount of information contained in a data set of this kind, and there are many aspects of it which the data analyst might consider. A methodical, step-by-step approach is advocated here, in which transformations of various kinds are used to assemble the information in the most appropriate frames of reference, and by removing redundant information, reveal the significant properties of the data more clearly. There is no unique path, however, by which all data can be analyzed, and the best techniques are determined by the nature of the data, the relations between variables, and the aim of the analysis.

3.8. Random variables, frequency distributions, descriptive statistics

The usual starting point with quantitative data is to consider each variable on its own, and examine the frequency of occurrence of the different values of the variable. The branch of mathematics known as probability theory indicates, on certain assumptions, the expected form of the frequency distribution. As a theoretical basis, the values of a variable in a sample are considered to have a predictable component resulting from the regular predetermined aspects of the underlying processes, on which is superimposed an unpredictable or 'random' component. The unpredictability may reflect lack of understanding by the geologist, as well as more fundamental limits to knowledge. A variable of this kind is known as a random variable or variate. Although the magnitude of the random component is, by definition, unknown for any specific item, the overall characteristics can be described. Some variables, such as the height of a human being, have a large unpredictable element, while others, such as the length of a part on a precision-made machine, vary within much narrower limits. The degree of unpredictability, or magnitude of the random element, is reflected in the variability of the variate. The population which is sampled by collecting a data set has a mean, average or expected value, of which it can be shown that the best estimate that can be obtained from the sample is the mean of the sample. The calculation of the mean is straightforward. The individual values are added together and the total is divided by the number of items.

In the ideal case of the so-called normal distribution, variation from the mean value is purely random, deviations are symmetrically distributed about the mean, the most frequently occurring value is the mean itself, and the likelihood of encountering a large deviation from the mean is less than that of encountering a smaller deviation. This is shown graphically in Fig. 35, where the familiar bell-shaped curve of the normal distribution indicates the likelihood of encountering values deviating to varying degrees from the mean value. From the sample of the population in a data set, a similar diagram can be prepared. As there is only a limited number of values in a data set, they are not usually depicted as a continuous curve, but as a bar chart or histogram. The full range of values in the sample is divided into

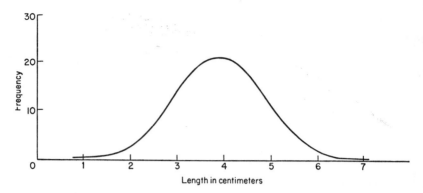

Fig. 35. A frequency distribution curve which is approximately the shape of the normal distribution curve.

segments, the number of values in each segment is counted and plotted in the vertical direction, as in Fig. 19. In terms of the geometrical representation of points in n-dimensional space, a number of layers have been defined by planes at right angles to an axis, and the number of points in each layer counted and plotted. The histogram gives a picture of the frequency distribution of a variate, including the position of the mean value, the amount of the variability, and the symmetry and regularity of the variation from the mean. After the mean, the most important aspect of the frequency distribution is the variability. This is measured by the variance, which is the average of the squares of deviations from the mean; or by the standard deviation, which is the square root of the variance.

3.9. Standardized data, error checking, data transformation

The mean and standard deviation measure properties which can be set in a larger context. The information that the average thickness of Jurassic strata in Ercu County is estimated to be 1003 feet with a standard deviation of 175 feet is information about the county as a whole, and can be used in making comparisons with other formations, or with the same formation in another county. It is also possible, from a knowledge of the mean and standard deviation, to transform individual measurements to show their magnitude in terms of the internal variation within the sample. The transformation is known as standardizing or normalizing the data, and consists of subtracting the mean from each of the original data values, and dividing the result by the standard deviation. The standardized data, therefore, show the number of standard deviations by which each item deviates from the variable mean. The calculation, which is quite straightforward, is shown in Fig. 36. Unlike the original measurements, the standardized data measure

A	B	C	D	E
Well Number	Sandstone thickness (feet)	Deviation from mean (feet)	Squared deviation	Standardized data
1	43	−5	25	−0·5
2	48	0	0	0·0
3	32	−16	256	−1·6
4	60	12	144	1·2
5	57	9	81	0·9
Totals 5 items	240	0	506	0·0
Average	48	Variance	101	
		Standard deviation	10	

Fig. 36. Standardization of data.

internal properties of the sample directly. The fact that a certain borehole has a standardized value of −2·5 for the thickness of the Jurassic, indicates that the formation is unusually thin within that group of wells. Standardizing has thus transformed the data from an external frame of reference, in which thicknesses are measured in feet, a unit that can be used to measure length in any context, to an internal frame of reference, where data are measured in the context of the properties of the sample. Another aspect of the standardized data is that both the mean and the standard deviation are measured in the same units as the original variable, namely, feet. The standardized values are calculated by dividing the mean (in feet) by the standard deviation (also in feet). The result is no longer measured in feet nor in any other unit, but is a dimensionless number. Thus variables, which originally measured very different properties of the sample, can be standardized, and thereafter compared on the same basis.

The first practical use that can be made of the standardized variables is to prepare a histogram of each variable for visual examination, see Fig. 37. The precise form of a normally distributed variate is known, and deviations from this can be readily spotted by eye. In computer analysis, the histogram is often prepared on a line printer at this stage in the analysis, since the remaining information is printed, and it is convenient, and saves time and expense, to have all the output prepared on one device. For publication, histograms can be prepared on a drum plotter. If all variates are standardized, then the histograms for each of the variables are directly comparable. They are likely to give useful information about the frequency distributions of the

RESULTS OF ROKDOC ANALYSIS

```
TITLE      LOUTO2
USER       T V LOUDON, SEDIMENTOLOGY, READING UNIVERSITY,
TITLE      SUMMARY OF LITHOLOGIES FROM FICTITIOUS BOREHOLES THROUGH
TITLE      THE DB6 ZONE OF THE CARBONIFEROUS OF THE GLARSTON-GLISWUG
TITLE      AREA OF CENTRAL SCOTLAND,
FORMAT     (6X,A4,2X,2F6,3,6F6,2)
FORMAT     (6X,A4,14X,6F6,2)
NAMVAR 00  BHOL  IDENTIFICATION NUMBER OF BOREHOLE              LIST OF VARIABLE NAMES
NAMVAR 03  THCK  TOTAL THICKNESS OF STRATA IN METERS
NAMVAR 04  NCYC  NUMBER OF COMPLETE CYCLES
NAMVAR 05  NCOL  NUMBER OF COAL SEAMS
NAMVAR 06  OTST  THICKNESS OF STRATA OTHER THAN COAL
NAMVAR 07  TRTY  TOTAL THICKNESS OF ROOTY BEDS IN METERS
NAMVAR 08  TSD   TOTAL THICKNESS OF SANDSTONE
NAMVAR 01  EAST  EASTING OF GRID REFERENCE
NAMVAR 02  NORT  NORTHING OF GRID REFERENCE
NUMERICAL INFORMATION FOLLOWS
NUMBER OF FIELDS       9,00000000
NUMBER OF FIELDS       7,00000000
```

*** THE NEXT TABLE LISTS THE RAW DATA ***

SMP NO		THCK	NCYC	NCOL	OTST	TRTY	TSD
1	SJ1	47.500	21.000	20.000	45.180	1.600	28.700
2	SJ2	76.500	37.000	32.000	72.270	2.300	16.200
3	SJ3	58.000	30.000	22.000	55.810	1.100	25.500
4	SE1	53.700	23.000	21.000	51.880	1.300	28.300
5	SE2	41.400	22.000	20.000	39.780	1.000	24.600
6	SE3	46.000	26.000	21.000	43.140	1.700	15.000
7	SE4	92.700	33.000	26.000	88.910	1.900	23.300
8	SE5	62.000	25.000	23.000	59.050	1.700	18.700
9	SE6	44.400	23.000	11.000	42.230	0.900	19.500
10	SE7	63.900	34.000	26.000	59.420	2.300	13.100
11	SW1	85.500	41.000	33.000	79.500	2.400	13.300
12	SW2	76.000	42.000	29.000	72.300	1.600	19.500
13	SW3	86.000	33.000	23.000	83.280	1.300	30.100
14	SW4	66.000	31.000	21.000	63.860	1.100	31.000
15	SW5	69.200	39.000	24.000	66.840	1.700	28.000
16	SW6	59.000	25.000	24.000	55.670	1.500	15.800
17	SW7	43.600	24.000	18.000	42.140	0.800	28.300
18	SW8	51.000	30.000	23.000	48.100	1.400	17.800
19	NJ1	55.500	31.000	21.000	52.800	1.700	19.600
20	NJ2	71.000	36.000	25.000	67.870	1.400	21.500
21	NJ3	64.600	31.000	19.000	62.340	0.900	27.200
22	NJ4	68.300	30.000	28.000	64.860	1.800	18.700
23	NJ5	83.600	37.000	27.000	80.160	2.300	23.200
24	NJ6	79.900	37.000	23.000	76.950	1.700	26.500
SMP NO		THCK	NCYC	NCOL	OTST	TRTY	TSD

THE FOLLOWING VARIABLES OF SUBSET HAVE BEEN STANDARDIZED TO HAVE ZERO MEAN AND UNIT VARIANCE
THCK NCYC NCOL OTST TRTY TSD

*** THE NEXT TABLE LISTS STANDARDIZED DATA ***

SMP NO		THCK	NCYC	NCOL	OTST	TRTY	TSD
1	SJ1	-1.156	-1.619	-0.731	-1.161	0.091	1.191
2	SJ2	0.829	1.004	1.900	0.774	1.625	-1.108
3	SJ3	-0.437	-0.143	-0.292	-0.402	-1.004	0.603
4	SE1	-0.732	-1.291	-0.512	-0.692	-0.566	1.118
5	SE2	-1.574	-1.455	-0.731	-1.547	-1.223	0.437
6	SE3	-1.259	-0.799	-0.512	-1.307	0.310	-1.329
7	SE4	1.938	0.348	0.585	1.963	0.749	0.198
8	SE5	-0.163	-0.963	-0.073	-0.170	0.310	-0.649
9	SE6	-1.368	-1.291	-2.704	-1.372	-1.443	-0.501
10	SE7	-0.033	0.512	0.585	-0.144	1.625	-1.679
11	SW1	1.445	1.660	2.119	1.291	1.844	-1.642
12	SW2	0.795	1.824	1.242	0.777	0.091	-0.501
13	SW3	1.479	0.348	-0.073	1.561	-0.566	1.449
14	SW4	0.110	0.020	-0.512	0.174	-1.004	1.614
15	SW5	0.329	1.332	0.146	0.386	0.310	1.063
16	SW6	-0.369	-0.936	0.146	-0.412	-0.128	-1.182
17	SW7	-1.423	-1.127	-1.169	-1.378	-1.662	1.118
18	SW8	-0.916	-0.143	-0.073	-0.952	-0.347	-0.814
19	NJ1	-0.608	0.020	-0.512	-0.617	0.310	-0.483
20	NJ2	0.453	0.840	0.365	0.460	-0.347	-0.133
21	NJ3	0.015	0.020	-0.950	0.065	-1.443	0.915
22	NJ4	0.268	-0.143	1.023	0.245	0.530	-0.649
23	NJ5	1.315	1.004	0.804	1.338	1.625	0.179
24	NJ6	1.062	1.004	-0.073	1.109	0.310	0.787
25	SUM	1545.299	741.000	560.000	1474.338	37.400	533.399
26	MEAN	64.387	30.875	23.333	61.431	1.558	22.225
27	STDV	14.609	6.099	4.561	13.996	0.456	5.435
SMP NO		THCK	NCYC	NCOL	OTST	TRTY	TSD

HISTOGRAMS FOLLOW, SHOWING THE FREQUENCY DISTRIBUTION OF EACH STANDARDIZED VARIABLE

VARIABLE 1 THCK

```
                                                          .
      .                                                   .
      .                                                   .
  5 -.                          +-+                        
      .                      +-+ +--+---1  +-+            :
      .                      1-+--+---1--1  1--1          :
      .                      1--1--1--1--+--+--1
      .                      +--+---1--1--1--1--+-+
  0 -. ##-##-##-##-##-##-##-##-##-+#+--1--1--1--1--1--1--1#+#-##-##-##-##-##-##-##-##-##-##-##-##-##- .
      -.....-.....-.....-.....-.....-....M-.....-.....-.-.
      -7    -6    -5    -4    -3    -2    -1     0     1     2
```

TOTAL N = 24
```

Fig. 37. Computer output from a program to calculate and print standardized data with histograms. Reproduced with permission from Loudon (1974).
```

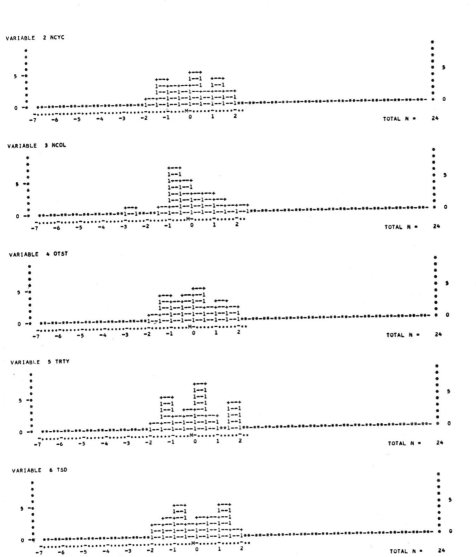

variables, the 'expected' bell-shaped curve of Fig. 35 being the exception rather than the rule. Where there is a major random component, a small sample from a normal population will not reproduce the ideal curve, and the histograms of samples of 10, 50 and 100 items from a normal population illustrated in Fig. 19, show how the irregularity of the frequency distribution increases as the size of the sample decreases.

Errors are inevitable at all stages of the data collection and recording process, and many of them can be picked up at this stage. Unusually high or low values, and values that do not 'look right' in the distribution are suspect. Such values should be examined carefully in a print-out of the standardized data matrix to determine whether they are associated with unusual values of other variates. The anomalous values in a distribution may well turn out to be the ones which carry the most geological or economic information. The first possibility to consider, however, is that they could be the result of data punched in the wrong column of a card, or transcribed wrongly on to a coding sheet. The second possibility is that the anomaly arose earlier in the data collection process by incorrect identification of the stratigraphy, or of a fossil species, by the wrong sample having been analyzed in error, by a soil sample for geochemical analysis having been collected near a discarded car battery, and so on. Computer analysis may indicate which samples are anomalous and require further study, but confirmation of the basic validity of the data is needed before geological conclusions are drawn. Another form of deviation from the ideal distribution may result from the sample having been collected from two or more populations with different parameters. Results of this are illustrated in Fig. 38. The detection of inhomogeneities is one purpose of cluster analysis, described below, but may also be suggested by the appearance of the histograms.

It is generally useful to have the computer print a table of standardized

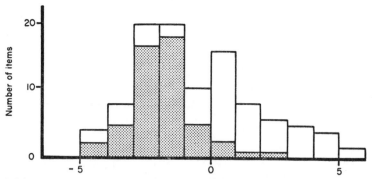

Fig. 38. Histogram of a mixed population. In a sample of 100 items, 50 are drawn from a normal population with a mean of 1·0 and a standard deviation of 2·50, and 50 (in the dotted area) from a normal population of mean −2·0 and standard deviation 1·25.

data. Individual values can then be compared to see whether exceptionally large or small values are accompanied by unusual values in other variates measured on the same item. General relationships may be apparent when the table is examined, such as the fact that groups of items with, say, slightly below-average Ca content tend to have above-average Na content, or that groups of items from one area or stratigraphic formation have properties that differ from those of other groups. The kind of properties that might be looked for are: unusually high or low values; more, or less, variation than in the sample as a whole; relationships between variates; and whether these are the same throughout the sample.

One of the more important reasons for examining the histogram is that the sample may be from a population which does not follow a normal distribution. The normal distribution occupies a central position in statistical theory, and is an assumption behind many of the statistical deductions that are made from probability theory. The emphasis in data analysis is on description and transformation, rather than deduction. Nevertheless, the statistical descriptors that are used to summarize the properties of a sample can be greatly affected by a small number of extreme values, even if they are not typical of the distribution as a whole. It is therefore desirable to transform the sample, if possible, into one which approximates to a normal distribution before descriptive statistics are calculated.

One transformation which is often helpful with geological data is the logarithmic transformation, which replaces each value with its logarithm. The valuable properties of the logarithm are well known from elementary algebra. The operation of adding the logarithms of two numbers together is equivalent to multiplying the two numbers. The operation of multiplying the logarithm of a number by a constant is equivalent to raising the number to the power of that constant. Thus multiplying the logarithm by two is equivalent to squaring the number. Many geological processes operate in a multiplicative, or so-called geometrical, manner, rather than an additive or arithmetic one. A fossil or a sand-dune, for example, may double in size with each year's growth, rather than becoming larger by a fixed amount. The degree of mineralization of a rock body may decrease exponentially with distance from the source of mineralization. The size of sedimentary particles which a stream can carry varies exponentially with stream velocity. It is perhaps not surprising, therefore, that many geological distributions are more nearly log–normal than normal. The consequences can be seen in histograms resembling the curve in Fig. 39, which show the typical shape of a log–normal frequency distribution. Further analysis of such data is simplified if the original data for that variate are replaced by logarithms of the values. The transformation can be applied in a routine manner in most computer packages for statistical analysis. In mechanical analyses of sediments, the

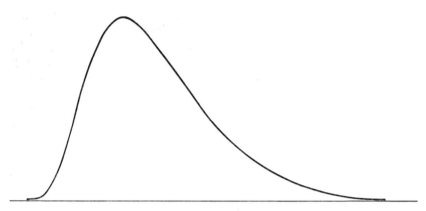

Fig. 39. An asymmetrical frequency distribution.

transformation may be applied before recording the data, by measuring sediment size in phi units, which is a logarithmic scale, and by using sieves which get larger in logarithmic steps.

A wide variety of other transformations can be applied to data in order to make them more amenable to analysis. With some data sets, the logarithm of $(x+1)$ is used rather than log x, to avoid difficulties arising from the logarithm of zero being minus infinity. The inverse, square, square root and trigonometric transformations are also used, though less frequently, in geology. They are described in detail, with advice on their use, in Miller and Kahn (1962). It is generally not advisable to make a transformation solely on the grounds that it gives the sample the appearance of having been drawn from a normal population without seeking some more fundamental explanation or justification in terms of underlying geological processes.

Data are often recorded with a somewhat higher precision than the instrument will support. A dip of 68° may be recorded from a clinometer that cannot be read accurately to less than five degrees, or semi-quantitative geochemical analyses may be recorded to a fairly high precision. Inspection of the histograms of such variates may show fluctuations due to this cause, which can be somewhat puzzling, particularly after transformation of the data. Each geologist has his 'favorite numbers' which appear more often than expected when the last figure is an estimate. This illustrates a general point that geological explanations of a pattern should not be sought until more obvious reasons have been considered and discarded.

After corrections, adjustments and transformations of the data set have been completed, the revised data set can be standardized and examined once more, and if all is well, the next steps can be taken in data analysis. In terms of the geometrical picture in n-dimensional metric space, the data points have been transformed so that they are symmetrically distributed about the origin, with a similar dispersion of points along each axis in each of the n dimensions, since each standardized variate has unit standard deviation. The effect of the transformations is shown in Fig. 40. In statistical terms, transformations have been found which make it possible on the one hand to summarize the properties of the distributions with a few parameters, and on the other hand, to arrive at an internal frame of reference which makes it possible to assess individual items and relationships within the sample. The data matrix, having been converted to this form, and with errors and spurious data points removed, can be handled more readily in the next stages of analysis.

3.10. Additional descriptive statistics; efficient and unbiased statistics

The mean and the standard deviation, which have already been described, measure two important properties of the frequency distribution. Other properties of the frequency distribution which are quite commonly measured are the skewness and the kurtosis. They can be calculated as the third and fourth root respectively of the average of the third and fourth powers respectively of the standardized variate. In practice, as with other statistics described later, the actual computation is likely to refer back to the raw data, to avoid carrying through rounding errors. The formulas for calculation then look more complex. Since the second and fourth powers of a number are positive, regardless of whether the original number is positive or negative, the standard deviation and kurtosis are always positive, and are not affected by the symmetry of the distribution about the mean. The skewness, on the other hand, is zero only if the distribution is evenly distributed on either side of the mean, and is greater than zero if the larger deviations from the mean tend to be positive, and less than zero if they tend to be negative. The fourth power is more strongly influenced by large deviations from the mean than the second power is, and hence the kurtosis is high if the distribution contains many large deviations from the mean. In general, the mean is a statistic which measures the location of a frequency distribution; the standard deviation measures its dispersion; the skewness its symmetry; and the kurtosis the shape of the distribution. Many other statistics could be calculated to describe properties of a frequency distribution, and indeed several are in everyday use in sedimentary petrography, designed for measuring from a cumulative frequency curve. The reason for statisticians generally tending to favor the four statistics just described is that they are efficient and unbiased. A statistic calculated from a sample is unlikely to correspond precisely with the population statistic, because of random variation. An

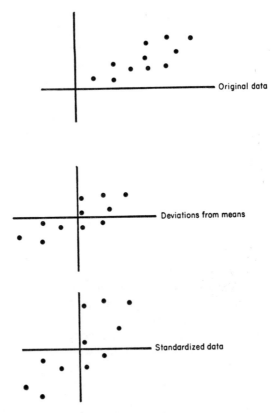

Fig. 40. Geometric representation of transformations of a data set, first to zero means, then to unit standard deviations.

efficient statistic, however, will tend to come closer to the true value than an inefficient one, while a biased statistic will tend to have values either consistently higher or lower than the population value.

3.11. Cluster analysis

Many geological variates are not measured from a single, homogeneous population. For instance, geochemical analyses might be made of rocks collected from two formations of different age and lithology, and it would not be surprising to find that the two groups had different frequency distributions and statistical values. However, the geologist is not always aware that he is sampling from a mixed population, and indeed, it may be one of the objectives of the data analysis to determine whether he is or not. Examination of the frequency distribution may suggest that the sample is bimodal or multimodal. The mode of a distribution is the value that occurs most frequently, and hence corresponds to the peak, or the highest bar on the histogram. In a perfect normal distribution, the mode and the mean coincide. In a bimodal distribution, such as is shown in Fig. 38, there are two separate peaks in the histogram, and in a multimodal distribution there may be even more. This may suggest that several populations with different properties are represented in the sample, or it may reflect a complex frequency distribution or may simply be a sampling effect. In either case, the existence of several groups, as in Fig. 41, is a possibility that may well require investigation. Cluster analysis provides a method of doing so.

In the example which was given above, a data matrix containing the position of a number of geological localities was analyzed to determine distances between every pair of outcrops. The distance matrix was then analyzed to find six clusters of adjacent outcrops, suitable for visiting on successive days of a field trip. The method adopted was to generate a dendrogram which displayed each locality in terms of distance from the nearest adjacent locality. The particular example was unrealistic, because the same results could have been achieved more effectively from a quick inspection of the points plotted on a map. Its value as an example lay in the ease of visualizing the procedure. Typical geological applications of cluster analysis involve many more than two dimensions, and may require more complex

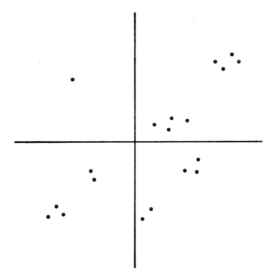

Fig. 41. A set of points clustered in space.

distance functions. The principles of the method, which will now be described, remain the same, and are more easily visualized in terms of the original example. The motives for undertaking cluster analysis are first, that it is desirable to know whether a sample can be regarded as having come from a homogeneous population, as this will determine what conclusions can be drawn from the analysis, and second, the existence of strong clustering in the data may reflect inhomogeneities which have geological significance, or might provide a means of classifying the population.

The sequential, aggregative method of cluster analysis can be applied to the data matrix, standardized as already described in order to give each variable an equal weighting. The first step is to calculate the distance between every pair of points, or items, to determine which is the closest pair. That pair is then linked, or regarded as a cluster, and the distance matrix is re-examined to find the next closest pair, which in turn are combined as the next cluster, and so on. Each successive linking is recorded on the dendrogram at the value of the distance function at which the linking occurred. The final stage links all items together as one large cluster. Items, once joined together, remain as a cluster for the remainder of the clustering process, and two main strategies are available for linking new items. Either the distance from the new item to the nearest point in a cluster is considered, or the distance from the new item to the centre of gravity of the cluster, that is, the average value of the cluster, is taken as the criterion for linking. The first method is known as single linkage, the second as average linkage. In both cases, a cluster, once formed, is treated as a point, and clusters are linked in the same way as points. Average linkage tends to encourage the development of 'spherical' clusters with approximately equal dimensions parallel to all axes. Single linkage allows unevenly shaped clusters to develop, and may thus help to detect groupings that would otherwise be missed.

Examination of the dendrogram in Fig. 34 also illustrates the fact that there is no single unique set of groups or clusters within a data matrix. The dendrogram is a tree structure which represents a hierarchy of groupings, and a classification could be established at any level of that hierarchy, from one extreme of including all the items in one class, to the opposite extreme of regarding each item as a class on its own. The differences between taxonomists who are 'lumpers' and those who are 'splitters' are reflected in cluster analysis. The most appropriate number of classes in a classification depends on the objectives, and the practical aspects of how many can be handled conveniently. The hierarchical nature of many classification schemes is reflected in the various divisions and subdivisions of rock types, stratigraphic units and fossils, such as family, genus and species. The value of a classification, however, may depend on the extent to which it recognizes or reflects real groupings where there is minimal variation within each class, and maximum distinction between classes. On the dendrogram, natural groupings of this type may be shown by a gap between linkages at a lower level, and the next higher set of linkages, see Fig. 34.

Various methods, many of which are numerically complex, have been devised to classify objects on the basis of characteristics for which a distance function can be defined, whether or not they are measured quantitatively. Sequential, aggregative cluster analysis, as described above, is a typical and widely used method. Applications have been in many fields, including bacteriology, archeology and textual analysis. In geology, typical problems to which cluster analysis is applied include: classification of microfossils; definition of sedimentary facies on the basis of the content of fossils and clastic grains in modern marine sediments; definition of lava types from chemical and petrological analyses of volcanic rocks; and classification of sedimentary environments on the basis of fossil assemblages. Cluster analysis is not, however, concerned with the problem of assigning new items to an existing classification, this being the domain of discriminant analysis. Cluster analysis may indicate that a data matrix can be divided naturally into several subpopulations, in which case, it may be advisable to return to the original data matrix, split it into these classes, and repeat the entire procedure of calculation of descriptive statistics, standardization of data, and cluster analysis separately for each class. Alternatively, the geologist may be satisfied that the sample is reasonably homogeneous, and that the analysis can proceed.

3.12. Correlation; principal component analysis

The next aspect to consider is the relationship between the variables in the data matrix. In geometrical terms, a relationship between two variables, x and y, can be illustrated in the x–y plot of Fig. 42, in which a marked tendency can be seen for high values of x to be associated with high values of y. If x and y are standardized variables, this tendency can be measured by the mean value of the product $x \cdot y$ of the standardized values. As before, to avoid rounding errors, or for desk calculator use, the value is calculated from the raw data, and a more complicated formula is used, in which the covariance, or average cross-product of the deviations from the means of the two variables, is divided by the product of their standard deviations. The result, in either case, is the correlation coefficient between x and y, for which the symbol r is often used. With one correlation coefficient between each pair of variables, a data matrix of n variates gives rise to an $n \times n$ matrix of correlation coefficients, sometimes termed the r matrix. Because the correlation between x and y is the same as that between y and x, the r matrix has the same values above and below the main diagonal, as can be

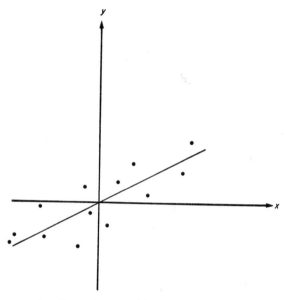

Fig. 42. An x–y plot, illustrating a positive correlation of two variates, x and y.

seen in Fig. 43. The entries on the main diagonal record the correlation of each variable with itself. This is the average square of the values of a standardized variable, namely the variance, or the square of the standard deviation, which, since the variate is standardized, is equal to $1 \cdot 0$. A value of $1 \cdot 0$ thus represents perfect correlation and is the highest value that the correlation coefficient can reach. The lowest possible value is $-1 \cdot 0$, since it is the mean cross-product of two standardized variables. This represents perfect inverse or negative correlation, and indicates that the values of x and y are again perfectly related but that as x increases, y decreases. A value of zero for the correlation coefficient indicates an absence of correlation, with the standardized data symmetrically distributed about the origin.

An examination of the correlation coefficient matrix indicates which pairs

```
***THE NEXT TABLE LISTS CORRELATION COEFFICIENTS BETWEEN VARIABLES***
```

VARIABLE	THCK	NCYC	NCOL	OTST	TRTY	TSD
1 THCK	1.0000	0.8002	0.7028	0.9985	0.5755	-0.0219
2 NCYC	0.8002	1.0000	0.7220	0.7894	0.5655	-0.1832
3 NCOL	0.7028	0.7220	1.0000	0.6755	0.7953	-0.4323
4 OTST	0.9985	0.7894	0.6755	1.0000	0.5409	0.0274
5 TRTY	0.5755	0.5655	0.7953	0.5409	1.0000	-0.5741
6 TSD	-0.0219	-0.1832	-0.4323	0.0274	-0.5741	1.0000
VARIABLE	THCK	NCYC	NCOL	OTST	TRTY	TSD

Fig. 43. Table of correlation coefficients calculated from the data of Fig. 37. Reproduced with permission from Loudon (1974).

of variates are most strongly correlated, and shows which are positively and which negatively correlated. Reference back to the standardized data matrix may throw some light on the nature of the correlation. A small data sample will always show some positive and negative correlations, because of the random element in the original data. Statistical tables give some guidance about the size of correlation coefficient that can be expected due to random variation alone, and this can help to indicate whether the correlations are significant in statistical terms. However, as is emphasized in the chapter on explanation, even if it can be shown that the r value is not likely to be the result of random sampling, geological explanation of the relationship may not be straightforward.

One unsatisfying feature of the correlation coefficient matrix is that many different relationships are measured in one table, and it is reasonable to suppose that several of the correlations may in fact be linked through a simpler set of underlying causes. If grain-size of a sediment is related to bed thickness and to percentage of quartz fragments and to the presence of broken brachiopod valves, it is more satisfactory to look for a single explanation, such as energy levels in the depositional environment, than to consider each pair of variates separately. Principal component analysis (PCA) is one of a number of multivariate statistical techniques, which by handling several variates simultaneously, provide an approach to the problem of interpreting complex relationships involving many variables. One objective of PCA is the parsimonious description of the complex relationships in a data matrix. Where twenty or more interrelated variables have been measured on every item of data, it is inevitable that the variates are to some extent measuring the same thing, and that much of the information they contain is therefore redundant. A more economical representation of the information may be achieved if new variables, known as principal components, can be created, which, although considerably fewer in number than the original variates, contain most of the same information.

The procedure is easiest to visualize geometrically. In the x–y plot of Fig. 42, it was seen how a distribution of data points in which x and y were correlated lay obliquely across the x and y axes. If both variates are normally distributed, the scatter will tend to be ellipsoidal. Two new axes can be selected, as shown in Fig. 44, one of which, x', lies in the direction in which the data points are most widely dispersed, and the other, y', is at right angles and therefore in the direction of least dispersion. Dispersion is measured by variance, and therefore if the original variates x and y are transformed, by rotating the axes to the new positions, to new variates x' and y', then the new variate x' will have a larger variance than x, and indeed will have the largest possible variance of any new variate that could be formed by rotation. The other new variate, y', will have the smallest

J

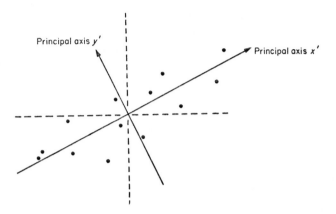

Fig. 44. The bivariate distribution of Fig. 42 referred to two principal axes.

possible variance. The total variance has not been changed, since the dispersal of points about the origin has not been changed. Because the original two variates were standardized, the total variance of x and y is $2 \cdot 0$, and this must also be the total variance of x' and y'. The variance of x', however, could be considerably more than $1 \cdot 0$, and of y' correspondingly less. It is also important to note that the distribution of points about the x' and y' axes is symmetrical, and that the correlation between them is zero.

Having located the new uncorrelated axes which account for the largest and smallest component of the variance, each of the original data points can be defined in terms of the new axes, and a new data matrix obtained by transforming the standardized data points to refer to the new axes. Geometrically, the operation is rotation. The geometric operation of rotation is conveniently represented in matrix algebra as matrix multiplication (see Pettofrezzo, 1969). The multiplication of two matrices is written $\mathbf{A}.\mathbf{B} = \mathbf{C}$, using the notation described in §2.5. The entry in the ith row and jth column of $\mathbf{C}$, c_{ij}, is the sum of the products of corresponding elements in the ith row of $\mathbf{A}$ and the jth column of $\mathbf{B}$. Thus, $c_{ij} = a_{i1}b_{1j} + a_{i2}b_{2j} + a_{i3}b_{3j} + \dots$. For matrix multiplication to be possible, the number of columns in $\mathbf{A}$ must equal the number of rows in $\mathbf{B}$ and columns in $\mathbf{C}$. The choice of a matrix to represent rotation is discussed in §3.13. The new axes are known as the principal axes of the distribution, and the new values as the principal components. No information has been lost, and none has been gained, in the transformation. Similar operations can be visualized in three dimensions, and algebraically, any number of dimensions can be handled. For example, if twenty variables had been measured in the original data matrix, twenty new variables could be found, of which the first accounted for the maximum possible variance, the second for the maximum possible variance of any axis uncorrelated with, or orthogonal to, the first, and so on. The last principal axis would be in the direction of minimum variance or dispersion for the entire distribution. The data could then be transformed to the twenty new principal axes, all of them uncorrelated or orthogonal, giving the value of twenty principal components for each item. Again, no information would have been gained and none lost. In geological terms, however, the situation has changed. Where previously twenty standardized variables had each accounted for one twentieth of the total variance, it is likely that the first three or four of the principal components will account for the greater part of the variance, or the total variability for which a geological explanation is sought. Furthermore, it is generally found that the significant pattern of the data matrix is shown by the first few principal components, while the remainder is largely random variation or noise.

The interpretation of the principal components is most readily understood in geometrical terms. For any item, the value of each principal component is measured as the distance of that item from the relevant principal axis. Each principal axis is oblique to some, and probably all, of the original axes. Its orientation is expressed in terms of these axes. If one thinks of a line, or vector, of unit length going from the origin along the principal axis (OC in Fig. 51), and perpendiculars are drawn from the end of that line to the original axes, then the lengths of the intercepts on the original axes are the coordinates of the principal axis, sometimes known as an eigenvector. The variance associated with the principal component is sometimes known as the eigenvalue. Algebraically, the value of the principal component for any item can be found by taking the values of the standardized variables, multiplying each by the appropriate coordinate, and adding them together. A matrix containing the values of all the principal components for all items can be obtained by a matrix multiplication of the standardized data matrix by the matrix of eigenvectors. It is not unusual to find that a direct physical significance can be ascribed to a principal component. For example, if the original variables were chemical elements, the principal components would consist of these elements mixed together in the proportions defined by the coordinates of the principal axes. These might represent a chemical compound, or a recognizable mixture of minerals. The values of the principal components, like those of the original variables, can be plotted on a map, and might, if appropriate, be contoured to look for pattern in their distribution in space. It is likely that only the first few principal components would show a recognizable geological pattern, the remainder representing largely random effects. An objective of PCA could thus be fulfilled, in that most of the pattern inherent in a large data matrix would be represented by a small number of synthetic variates. The extensive redundancy of the original data is considerably reduced. Furthermore, since PCA may reveal the extent to which several variables are in effect measuring the same thing, it may suggest methods of more economical data collection. Those variables could be selected for measurement in future studies which throw most light on the geological relationships as shown by PCA, and which can be cheaply and quickly measured and recorded.

There are related techniques, often considered together with PCA under the general heading of factor analysis, which may in certain cases give results that are more easily interpreted than those of PCA. They tend, however, to make assumptions about the distribution of variables in the original data, and thus do not have the same generality. Unlike most of the other methods of factor analysis, PCA can be regarded as a purely descriptive technique. If it is viewed geometrically as the rotation of a multidimensional distribution of data points, it can be seen that this calls for no assumptions about the original distributions, although, for reasons which were discussed previously, the method is likely to reflect the overall aspects of the data matrix more effectively if the distributions are standardized and nearly normal.

The procedures described above are examples of transformations that can be used in data analysis in an attempt to reveal geologically significant aspects of the data in an appropriate frame of reference and separated from redundant information. Other multivariate methods, which are not described in detail here, include non-linear mapping, a transformation applied to the data matrix in which as much information as possible about the relationships between variables is retained while it is reduced to two dimensions for display as a diagram, and discriminant analysis, in which a set of measurements on an item is transformed into a measure which guides the assignment of a new item to the correct class in a known, pre-existing classification, see Davis (1973).

3.13. Geometric data processing

The analogies between the algebraic methods of data analysis and their geometric equivalents have already been stressed. The same algebraic techniques are relevant to the direct geometric manipulation of geological data. The basic fundamentals of geometric data processing are outlined in order to emphasize the breadth of application of numerical techniques, and the extent to which a common solution can be found, by suitable transformations, to problems which initially may appear to be unrelated. Geometric data processing has not been widely discussed in the geological literature, but there can be little doubt of its fundamental importance, nor of its growing application in fields such as automated cartography. The requirement arises from the three-dimensional distribution of most geological data, and the importance of pattern in space in interpreting the data. Computer-drawn maps, cross-sections and perspective views can assist the geologist to display the data in the most effective way, and the elements and relationships that are manipulated on paper by the geologist, the draftsman and the cartographer in preparing maps and diagrams can also be handled within the computer, and thus linked with other interpretative methods. The basic geometric transformations, which are described below, are: translation, or shift of origin; rotation; stretching and change of scale; shearing; reflection; and projection.

The framework of coordinate geometry in which geometrical data are handled in the computer has already been described (see Fig. 51). Three lines at right angles, conventionally directed east, north and vertically upwards, form the three rectangular or orthogonal axes, x, y and z. The axes meet at the origin, conventionally labelled O on diagrams. The position of a point in this space is defined by its coordinates, which are the distances along the axes from the origin to the point at which a perpendicular from the data point meets the axis. The coordinates are written as a sequence of letters or numbers, one for each axis, separated by commas and enclosed in brackets. Thus the coordinates $(3 \cdot 0, -2 \cdot 6, 1 \cdot 4)$ would refer to a point $3 \cdot 0$ units east of the origin; $-2 \cdot 6$ units north, that is to say $2 \cdot 6$ units south, of the origin; and $1 \cdot 4$ units above the origin. The positive directions are east, north and upwards in this case, and a negative value thus indicates movement or distance measured in the opposite sense, namely west, south or downwards. A set of locations in two dimensions can conveniently be represented by two columns in a data matrix, and in three dimensions by three columns.

Movement of data points relative to the origin is known as translation, the word being used with quite a different meaning to that of dictionary lookup mentioned before. All movement is relative, and translation of the data points relative to the origin has the same effect as movement of the origin relative to the data points. Geometrically, the operation of translation is illustrated in Fig. 45. Algebraically, as performed on the computer, the operation consists of adding the appropriate number of units to the coordinates of all relative points. For example, if the data points were to be moved 2 units towards the left-hand side of a map, and 1 unit nearer the top, the translation $(-2 \cdot 0, 1 \cdot 0)$ would be performed by adding $-2 \cdot 0$ to all x values in the data matrix, and $1 \cdot 0$ to all the y values. This is the same as moving the origin by $(2 \cdot 0, -1 \cdot 0)$.

Rotation is an angular movement of all data points about the origin, the rotation being about one axis, or about several axes in turn. The order in which a sequence of rotations or other geometric operations is performed makes a difference to the result, as simple experiments confirm. The geometrical operation of rotation is illustrated in Fig. 46. The algebraic equivalent is matrix multiplication of the coordinates of each point by a

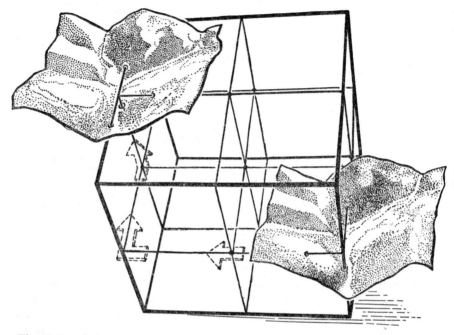

Fig. 45. Translation, or movement relative to the origin, of an area from which a set of samples have been collected.

2×2 matrix for a two-dimensional rotation, and a 3×3 matrix for a three-dimensional rotation. This is explained most easily by illustration. Fig. 47 shows a point P, with coordinates (x_0, y_0, z). After rotation through an angle θ about the z axis, the new coordinates are $(x_0 \cos \theta - y_0 \sin \theta, x_0 \sin \theta + y_0 \cos \theta, z)$. Multiplication of the coordinates (x, y, z) by the matrix

$$
\begin{bmatrix}
\cos \theta & \sin \theta & 0 \\
-\sin \theta & \cos \theta & 0 \\
0 & 0 & 1
\end{bmatrix}
$$

is the equivalent to rotation through an angle θ about the z axis. Similarly, multiplication by

$$
\begin{bmatrix}
1 & 0 & 0 \\
0 & \cos \phi & \sin \phi \\
0 & -\sin \phi & \cos \phi
\end{bmatrix}
$$

is equivalent to a rotation ϕ about the about the x axis. Just as the order in

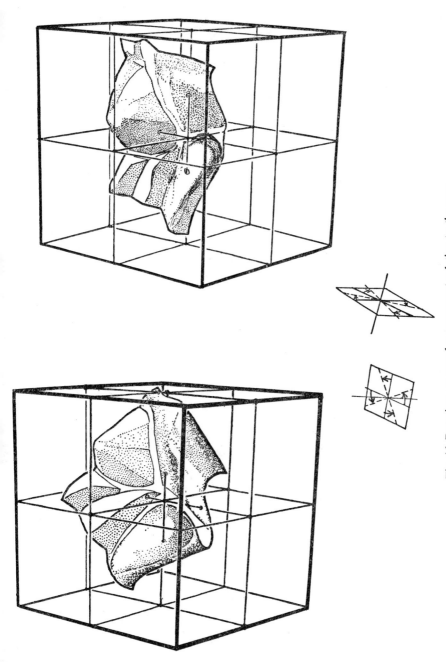

Fig 46. Rotation, or angular movement relative to the axes.

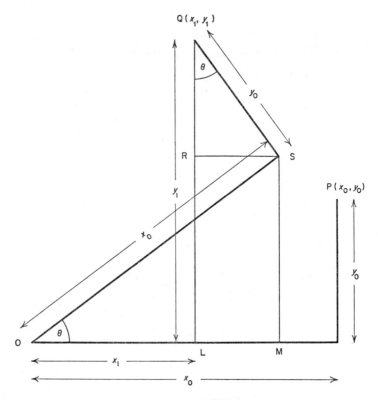

Fig. 47. Rotation takes the point (x_0, y_0) to (x_1, y_1), where $x_1 = x_0 \cos \theta - y_0 \sin \theta$; $y_1 = x_0 \sin \theta + y_0 \cos \theta$.

which the rotations are performed is important, so the matrix multiplications must be performed in the correct order. However, a sequence of rotations can be combined in one matrix, by multiplying the matrices representing the component rotations together in the correct order. Clearly, a shift of origin prior to rotation has a different effect to the same operations performed in reverse order. Rotation of a set of points can be handled by storing the data set as a matrix, and multiplying it by the rotation matrix, thus performing one matrix multiplication, rather than a series of multiplications of vectors by the matrix.

The effect of change of scale is illustrated in Fig. 48. The scale may be changed to the same extent parallel to all axes, or the scale on one axis may be changed relative to the other two, with the effect of stretching or elong-ating objects parallel to that axis, as though a diagram drawn on rubber was being pulled apart. Change of scale can again be represented by matrix

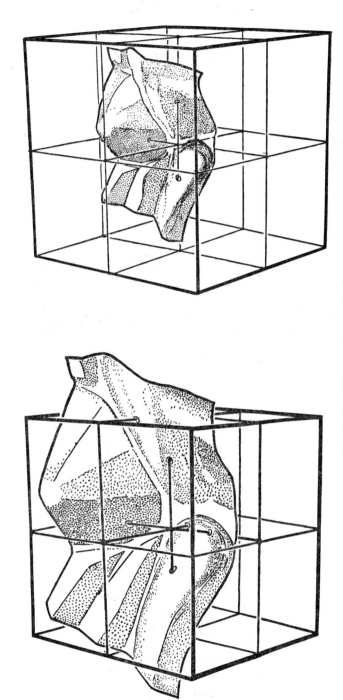

Fig. 48. Stretching, or change of scale.

multiplication. The matrix

$$\begin{bmatrix} 1 & 0 & 0 \\ 0 & 1 & 0 \\ 0 & 0 & 1 \end{bmatrix}$$

is the identity matrix, which on multiplication, leaves all points as they were. The matrix

$$\begin{bmatrix} 2 & 0 & 0 \\ 0 & 0{\cdot}5 & 0 \\ 0 & 0 & 1{\cdot}0 \end{bmatrix}$$

would have the effect of increasing all x values by a factor of 2, halving all y values, and leaving z unaltered. The effect of multiplying a set of coordinates by that matrix would be to compress the area to half its previous dimensions in the north–south direction, and stretch it to twice its previous size in the east–west direction, leaving vertical distances unchanged.

The operation of shearing is one which causes a rectangle to become an oblique parallelogram, as shown in Fig. 49. It is less geologically important than the other geometric operations, although it has relevance in tectonics and geotechnics.

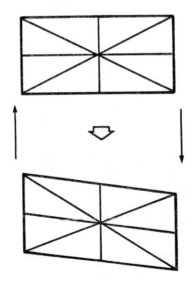

Fig. 49. Shearing.

The matrix representation is

$$\begin{bmatrix} 1 & 0 & 0 \\ \tan\theta & 1 & 0 \\ 0 & 0 & 1 \end{bmatrix}$$

where θ is the angle of shear, measured from the y axis. Reflection is an operation that may be familiar from crystallography. A mirror image of the data points with respect to the z-axis can be obtained by multiplying them by the matrix.

$$\begin{bmatrix} 1 & 0 & 0 \\ 0 & 1 & 0 \\ 0 & 0 & -1 \end{bmatrix}.$$

The basic operations mentioned above can be combined by setting out a sequence of operations in matrix form, and then multiplying them together to give one matrix which represents the composite transformation. Stretching in a line oblique to the axis can thus be achieved by rotating the distribution so that an axis lies in that direction, then stretching, and finally rotating back to the original position. Translation, which involves addition rather than multiplication, cannot be included in this way in a two-dimensional matrix.

The geometric operations of translation, rotation, shearing, reflection and change of scale, can be brought together as one composite matrix operation, known as an affine transformation. It is a transformation of location data from one metric to another, in which straight lines remain straight, parallel lines remain parallel, and the ratios of lengths of segments of a line remain the same. The use of affine transformations can be extended by applying them to homogeneous coordinates, which represent a point in two dimensions by a three-dimensional vector from the origin passing through the point. The point (x, y) has homogeneous coordinates (x, y, z) where the z value is somewhat arbitrary. The three values can be multiplied by a constant without moving the point from the vector. The effect of multiplying by 2, to alter all z values from say, $1\cdot0$ to $2\cdot0$, can be visualized, as in Fig. 50, as a projection of data from a transparency to a screen. If the transparency were at an angle to the screen, the effect of the transformation would be to alter the scale of x systematically in the y direction and vice versa. This, as explained by Ahuja and Coons (1968), is one method of transforming from, say, an irregular quadrilateral to a rectangle, and thus of correcting certain kinds of distortion in a map or diagram. The more complex transformations required for changes from one map projection to another can be performed by computer, but

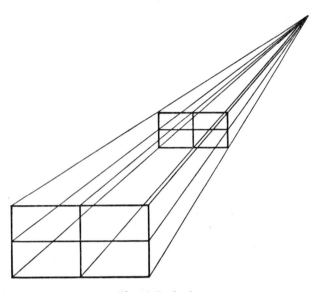

Fig. 50. Projection.

but are a rather specialized aspect of cartography. When a program package is used, the mechanics of the operations may be handled automatically, and the user may simply have to list the series of elementary operations that have to be performed.

Geometric data processing is an integral part of the preparation of computer plots. Data derived from various maps and photographs on different scales and for different areas may have to be brought together for display on a single plot. This may mean that individual sets of data have to be manipulated before they can be presented on the same basis on the final plot. This is the case, in particular, where the original locations of the data points are recorded by automatic digitizing which records the relative positions of points on the digitizing table together with information which makes it possible to convert from the table coordinates to the map coordinates, as well as correcting for internal distortion of the digitized material, due, for example, to paper shrinkage. Geometric data processing also has an application of more immediate relevance in data interpretation. This arises from the fact that the geographic space in which the data were originally recorded is not necessarily the most effective space for data interpretation. The concepts of geometric data processing make it possible to transform data sets to bring patterns in space to a common form and to make clearer the similarities between apparently different patterns. If, for example, the geometric features of a number of sand bodies had to be compared, a sequence of operations could alter the frame of reference, while identifying, measuring and eliminating aspects of the geometry. Location, for instance, could be measured for each sand body, as the coordinates of its centre of gravity, subsequent measurements taking the centre of gravity as the origin. The sizes of each body could be made comparable by change of scale. They could be brought to comparable orientations by appropriate rotations to principal axes. Symmetry could be achieved within each sand body by appropriate stretching and compression parallel to the principal axes. At each of these steps, a measurement is obtained of that particular property for each body. At each step, the remaining differences are placed within a frame of reference that makes subsequent comparison easier. The approach is analogous to that of calculating the mean, standard deviation, correlations and principal components of quantitative data and recomputing the original data values to refer to the new frame of reference.

3.14. Analysis of orientation data

Orientation data, such as the strike and dip of bedding planes or the orientation of cross-bedding, are frequently recorded in geology. The methods of data analysis described above cannot be applied directly, because, being measured in angles rather than coordinates, the data are in a different metric space. A suitable transformation can, however, bring orientation data into a suitable frame of reference. Orientations measured in polar coordinates, such as strike and dip, can be transformed to direction cosines, one form in which orientation data are handled in coordinate geometry. The direction cosines (l, m, n) describe the orientation of a vector in three dimensions, by three numbers between $-1 \cdot 0$ and $1 \cdot 0$, written in a similar way to geographical coordinates, namely, enclosed in brackets and separated by commas. The three numbers are the cosines of the angles between the vector and each of the three rectangular coordinate axes. A vector in geometrical terms is a line with a defined length, orientation and sense. As shown in Fig. 51, if a vector of unit length is directed away from the origin, the lengths of the perpendiculars to the axes from the end of the vector are equal to the direction cosines of the vector. A line directed towards the negative end of any axis, has a negative value for that cosine. From the same diagram, it can be seen that, because of the theorem of Pythagoras, $l^2 + m^2 + n^2 = 1$.

An advantage of direction cosines as a means of representing orientation is that they can be handled in a similar way to coordinate data. Direction cosines can be rotated, stretched and skewed using the same matrix operations as apply to coordinate data. The vector mean can be calculated for a set of vectors, by calculating the mean values of the set of l values, and of m and of n values. The three mean values give the direction ratios of the

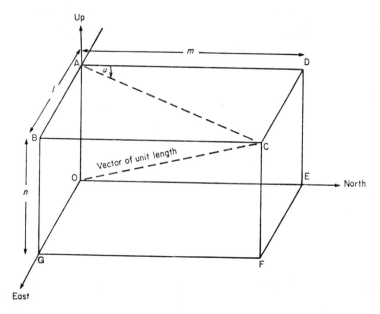

By definition, the direction cosines (l, m, n) of the unit vector OC are: $l = \cos$ GÔC
$=$ OG $=$ DC; $m = \cos$ EÔC $=$ OE $=$ AD; $n = \cos$ AÔC $=$ OA. If the unit
vector OC is the pole to a plane which dips at v degrees in a direction u degrees
east of north, then DÂC $= u$, and AÔC $= v$.

$$AC = \sin \text{AÔC} = \sin v, \qquad l/\sin v = DC/AC = \sin DAC = \sin u.$$

Hence, $l = \sin u . \sin v$

$$m/\sin v = AD/AC = \cos \text{DÂC} = \cos u.$$

Hence, $m = \cos u . \sin v$

$$n = \cos \text{AÔC} = \cos v.$$

Fig. 51. Direction cosines (l, m, n) describing the orientation of a vector in three dimensions.

vector mean, rather than the direction cosines, because the vector mean
is of less than unit length, and the sum of squares of the direction ratios
is less than one. The length of the mean vector can in fact be calculated as
the root of the sum of squares of the direction ratios, and it has been proposed,
by R. A. Fisher (1953) as a measure of the variability of the set of orientations.
Direction ratios are easily converted to direction cosines by dividing the
three values by the vector length, and can then be converted back into strike
and dip or other polar coordinates. In structural geology, the frequency
distribution of orientations has rather different properties when considered
parallel to and normal to the main axes of folding. Using the same methods

as PCA, principal axes to the distribution of vectors can be found, one of which corresponds to the fold axis, and the eigenvalues, which give the variance associated with each axis in PCA, are measures of the intensity of folding in the two directions (Loudon, 1964; Watson, 1966).

3.15. Regression; trend surfaces

The type of relationship which has been described so far concerns several variables that play an equal and similar role in the relationship, and the variables are considered together in multi-dimensional space. However, this does not necessarily represent the model which the geologist has in mind. Many geological data refer to a cause and effect or process and response model, where one variable can be clearly recognised as describing an aspect of the phenomenon which the other variables may, in some sense, explain. For instance, the value of, say, uranium content in a black shale might be related to its location, that is, the east and north variables. The mean grain size of a sediment might be related to distance from source and energy gradient across the depositional environment. One variable is of primary interest, and the others play some part in affecting or controlling its value. This may be regarded as a tentative model, which the geologist tests on several groups of variables in turn, or as a firmly established model for which data are collected to establish accurate values for the model parameters. A mathematical and statistical framework, misleadingly named regression analysis, is well established for handling this class of problem.

The mathematical model of regression analysis is written as an equation in which a single variable, say y, appears on the left-hand side, and one or more variables, say x_1, x_2, etc. appear on the right-hand side, each preceded by a coefficient, a, b, c, etc. In addition, an error term (ε) appears on the right-hand side, thus: $y = a + bx_1 + cx_2 + \ldots + \varepsilon$. In the examples given above y might be parts per million of uranium; x_1 the easting of the locality reference, and x_2 the northing. The equation would then represent a model in which the uranium content can be predicted or approximated from the other two variables by taking the value of the constant, a, adding to it the value of the easting multiplied by b, and of the northing multiplied by c. Regression analysis of the data matrix is the method of estimating numerical values of a, b and c from the sample. The presence of the error term ε indicates that the calculated value y is subject to error or random variation, and therefore the calculated value will not normally be identical to the measured or observed value. An example of a simple case with only two variables may help to clarify the procedure.

With two variables, the regression equation may be more familiar as $y = mx + c$, a relationship illustrated in Fig. 52, already introduced as an

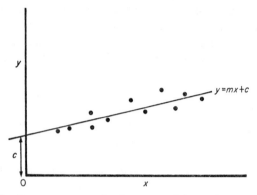

Fig. 52. A regression line, $y = mx + c$, fitted to a set of data points (x, y) shown by dots.

x–y plot. The values of x and y for data items in the sample are shown as dots in the diagram. The regression line is the line which passes most closely to all the data points. The coefficient m is a measure of the slope, or gradient of the line, and c is a constant which indicates the height of the line in the diagram, in terms of the intercept of the line with the y-axis. If x and y were standardized variables, it might be expected that c would equal $0 \cdot 0$, with the line passing through the origin, and the slope m would be an indication of the manner in which the two variables were related, akin to the correlation coefficient r. However, unlike the correlation coefficient and PCA models, in which variables are interchangeable and are all regarded as random variates, in regression analysis one variable, y in this case, has a special significance, with the coefficients of all the other variables in the regression equation being selected to predict the value of y as closely as possible. The error term reflects the impossibility of predicting a random variate with complete certainty. In multiple regression, there is no theoretical limit to the number of variables that can be considered on the right-hand side of the equation. It is not, however, a multivariate method. A variate is a random variable, and in regression only one error term appears, implying that only the variable on the left-hand side of the equation is regarded as random.

Principal component analysis, applied to the variables x and y of Fig. 52 would attempt to fit principal axes through the data points in a symmetrical manner, so that the variances about the principal axes were a maximum and a minimum. The sum of squares is therefore minimized perpendicular to the best-fit axis. In regression analysis, however, the error term applies to the y value and the best-fit regression line is chosen to minimize the sum of squares of distances measured in the y-direction, from the data points to the line. In multiple regression, coefficients are calculated for all the terms on the right-hand side of the equation, so that, for the items in the sample, the sum of squares of deviations of the measured values of y from the values calculated from the regression equation is a minimum. The contribution of each variable is not independent, unless the variables themselves are uncorrelated (that is, orthogonal), and if some of the variables are dropped, the entire regression equation must be recalculated.

A straight line can be fitted to a set of data points by simple regression. A curved line or surface can be fitted by multiple regression, if an appropriate series of terms is selected. This aspect is particularly important in geology. The equation of a straight line can be written $y = mx + c$, or $y = a + bx$; a quadratic curve $y = a + bx + cx^2$; a cubic curve $y = a + bx + cx^2 + dx^3$; and so on. Typical curves are illustrated in Fig. 53. In general, by adding further terms to the power series $a + bx + cx^2 + dx^3 + \ldots$ a curve of increasing complexity and with an increasing number of turning points can be defined to approximate as closely as required to the data points. In practice, since there is a random element in the y variate, an exact fit is seldom the objective—fortunately, because the number of terms required might equal the number of data points. As close a fit as is thought to be useful can be obtained by sequential regression, in which curves are fitted with an increasing number of terms, adding one more at each iteration, until a satisfactory fit is obtained. This method of polynomial regression is a special case of multiple regression, in which the variables are the successive members of a power series.

The other type of series that is important in curve-fitting is the trigonometric series. Successive terms representing sine and cosine waves of different wavelength and amplitude make up the right-hand side of the equation. Like the power series, the trigonometric series has the valuable property that as more terms are added, the series approximates increasingly closely to the required curve. An important difference between the two is that the

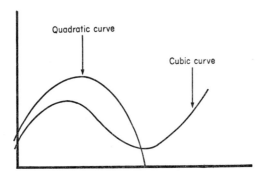

Fig. 53. Examples of the approximate shape of quadratic and cubic curves, which can be fitted to data by multiple regression.

K

sine waves produce a curve which repeats the same shape indefinitely as it is extended beyond the data points, whereas the polynomial curve either continues in a straight line or curves asymptotically towards an axis.

In both the above examples, x and y are two variables which have been measured and recorded in a data set. It is entirely possible, however, to have a regression equation linking y to a polynomial in two variables, say x_1 and x_2. The regression equation then has the following form:

$$y = a + bx_1 + cx_2 + dx_1^2 + ex_1x_2 + fx_2^2 + gx_1^3 + hx_1^2x_2 + ix_1x_2^2 + jx_2^3 + kx_1^4 + \ldots$$

which can be extended to any required number of terms. The most important application of such an equation in geology is in trend-surface analysis, where x_1 and x_2 are geographical coordinates such as eastings and northings, which measure position in space. The geological variable, y, could be any variable that can be represented as a surface by a contour or isopach map. The number of terms included on the right-hand side determines the complexity of the surface represented. A cubic surface, which includes terms to the power 3, gives a relatively smooth surface which may nevertheless reveal a pattern which has geological significance. It is at a suitable level of complexity for representing a typical geological trend surface.

The regression equation thus represents a line or surface in multidimensional space which best approximates to the distribution of data points in that space. The best approximation is defined to be that equation which minimizes the sum of squares of deviations from the y variable, measured parallel to the y-axis. The purpose of computing such a function may be to obtain an overall representation of a relationship, giving a picture of the behavior of the variables within the population as a whole, rather than just within the sample. This in turn may be of value in interpolating or predicting values of y where the variables on the right-hand side are known. The value to the geologist may be in the summary of the relationships, and comparison of these with similar summaries for other areas or other circumstances, or for checking against existing ideas of how the variables are related. But, as with other summaries, such as the mean and standard deviation, the regression equation provides a local frame of reference against which variation within the sample can be examined. Deviations of data points from the regression line or surface can be the most informative part of the regression analysis. It is interesting, therefore, to consider what types of deviation might be expected and how they could arise.

An obvious, but sometimes overlooked, starting point in looking at deviations from the regression equation is to consider whether the regression is in fact reflecting a real pattern in the data, or whether it is merely a best fit to a scatter of random points. Computer programs for regression analysis usually indicate the proportion of the total variance of y which is accounted for by the regression, and the proportion which remains unexplained. The greater the extent to which the regression equation accounts for the variance, the more likely is it to be reflecting a genuine pattern. If a number of assumptions are fulfilled, then statistical tests are available to help to assess this, as mentioned in the chapter on explanation. With sequential multiple regression, it is possible to measure the separate contributions of the individual terms on the right-hand side of the regression equation, by the amount of total variance of y accounted for by each.

If there is reason to accept the regression equation, the pattern of individual deviations of data points can be examined, as an x–y plot (Fig. 52) for a line, and a contoured map of deviations for a surface. The regression line is a mathematical artefact which is not necessarily the best representation of the relationship. A point to examine in the deviations, therefore, is whether they show a pattern due to the regression rather than to the data. Fitting a sine function, for instance, may create periodic deviations because periodicity in one part of the line or surface is carried through to data where the periodic element is no longer present, or is out of phase. Polynomial curves may be of the wrong order to fit the data adequately, and in consequence, the deviations may themselves show a strong pattern when inspected visually. If a cubic trend surface is fitted to a sedimentary basin which has a curved axis, as in Fig. 54, the three-dimensional distortion will not be met by the trend surface, which may therefore show deviations on one side of the axis at each end of the basin, and on the other side at the centre. A simple geological shape is not necessarily a simple mathematical shape, and it is important to keep in mind the properties of both and the possibility of a mismatch between them.

After more obvious possibilities are ruled out, the geological reasons for the deviations can be looked at, by comparing them with the values in the data matrix. The deviations might be related to another variable which was not included in the regression, and possibly not even measured, or there might be a non-linear relationship with one of the existing variables, so that it was not fully defined by simple regression. The deviations, when plotted on a map, may show a pattern in space. For instance, if the grain-size of a sampled sediment was largely accounted for by a regression model in which distance from source and energy gradient of the depositional environment were included as variables, and if deviations between the observed grain size and values calculated from the regression equation were plotted on a map, areas of generally high and generally low deviations might indicate secondary sources of the sediment. In trend-surface analysis, the regression surface is likely to reflect large-scale geological processes, such as regional subsidence or folding, while deviations from the surface may reflect relatively small-

Curved axis of   Axis of cubic
sedimentary basin   trend surface

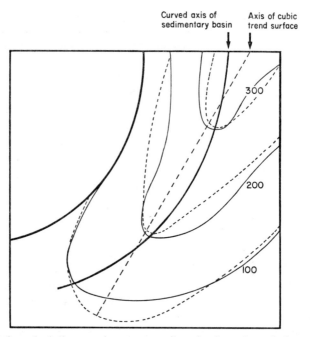

Fig. 54. A simple geological pattern is not necessarily a simple mathematical one. The solid lines are contours on a basin with a curved axis. The dashed lines show the approximate position of a cubic trend surface fitted to the data. Residuals would require cautious interpretation.

scale or local effects, particularly those of irregular form, like river valleys, reefs or salt domes. The regression model can, in general, be a useful tool for separating out a component of variation attributable to a widespread and defined model, from a local effect which requires separate explanation. It is important for the geologist to know of local anomalies, as they could well have economic or scientific significance, which would justify the geologist concentrating his attention on them.

3.16. Analysis of qualitative data

Correlation, principal component and regression analysis are techniques which can be applied to a set of quantitative variables drawn from a reasonably homogeneous population. In a geological investigation, a homogeneous population is the exception rather than the rule. Where qualitative as well as quantitative variables are included in the data analysis, the qualitative variables typically subdivide the sample on some basis, such as type of lithology, formation or facies. The dataset may thus be divided into several distinct subsets, or may be divisible on the basis of several over-lapping criteria, forming different subsets depending on which criterion is used.

The methods already described can of course be used within a homogeneous subset, but the use of PCA or regression analysis on a sample drawn from several subsets would tend to reflect the proportions in which the various subsets were represented in the sample. The objective of analyzing categorized data may be to establish the manner in which the properties vary from one category to another, or to determine the extent to which the various sets of categories interact with one another and with other variables. An answer might be sought to such questions as: How is the copper content of the Paleozoic sediments of Ercu County affected by the lithology, the stratigraphic level, the color of the sediment, and the geographic location? Is the color of the sediment related more strongly to lithology, environment of deposition, or age of rock? The methods of analysis that are appropriate depend largely on the type of data available. If the data are entirely qualitative and the number of categories or classes is not too large, then a useful summary of the relationships is provided by the so-called contingency table. The categories form the rows and the columns, and the entries show the number of simultaneous occurrences of the row-category and the column-category. For example, if the data matrix referred to a succession of stratigraphic units encountered in a borehole, the matrix might show the number of units that were sandstone and brown, limestone and brown, sandstone and grey and so on, as in Fig. 55. The marginal totals show the number of units with each characteristic. The matrix showing frequency of co-occurrences may itself be of considerable interest. The total number of sandstone beds and of grey units, however, determines the expected frequency of grey sandstone units. The point of interest is likely to be whether there is a tendency for the qualities of sandstone-ness and greyness to be associated or otherwise. Does the fact that a unit in the data set is sandstone make it more or less likely to be grey than would otherwise be the case? A comparison with the values that would be expected if the attributes were independent may throw light on this question, see Fig. 55.

A convenient tabular summary which includes both qualitative and quantitative data can be obtained by combining the contingency and the correlation coefficient tables. Where the row and the column both refer to quantitative data, the entry is the correlation coefficient between the variables. Where both row and column are qualitative variables, the entry is the probability of co-occurrence. Where the entry refers to one qualitative and one quantitative variable, the result is the mean value of the quantitative variable for items having the indicated attribute. It is thus possible, by scanning the table, to compare the mean value of a variable for the entire data set with

(a)	Color				
Lithology	Red	Grey	Brown	Black	Totals
Conglomerate	0	4	6	0	10
Sandstone	6	6	8	0	20
Siltstone	0	15	5	0	20
Shale	4	15	11	10	40
Limestone	0	10	0	0	10
Totals	10	50	30	10	100

(b)	Color				
Lithology	Red	Grey	Brown	Black	Totals
Conglomerate	1	5	3	1	10
Sandstone	2	10	6	2	20
Siltstone	2	10	6	2	20
Shale	4	20	12	4	40
Limestone	1	5	3	1	10
Totals	10	50	30	10	100

(c)	Color			
Lithology	Red	Grey	Brown	Black
Conglomerate	1·0	0·2	3·0	1·0
Sandstone	8·0	1·6	0·7	2·0
Siltstone	2·0	2·5	0·2	2·0
Shale	0·0	1·2	0·1	9·0
Limestone	1·0	5·0	3·0	1·0

Fig. 55. Contingency tables. (a) Observed frequencies (O) of occurrence of beds of a given lithology and colour. (b) Expected frequencies (E) if lithology and color were not related. (c) A measure $(O-E)^2/E$ of the deviations between the observed and expected values.

those for selected subsets of the data, and to determine, for instance, how the uranium content of a shale varies from one color class to another, and varies from one area to another. Although this can in itself be informative, it may also be necessary to consider the number of items and the variability within each category, to determine whether the differences are significant, and which qualitative variables play the biggest part in controlling the variation. The variability is of course measured by the variance or by the standard deviation. The analysis of variance is a major study in itself, and its application to geological explanation is considered in the next chapter. For interpretation of data, it is relevant to consider how information on means and variances within categories can be organized and displayed in the most effective manner.

Geological data can in many cases be divided into categories on the basis of more than one criterion. For instance, geochemical data may refer to samples which are classified by stratigraphy, lithology, geographical area and type of sample. It may be informative to know what the average uranium content and the variance are for each lithological group, as a guide to the lithologies most likely to contain economic deposits. Both the mean and the variance have a bearing on this, since a high value for both the mean and the variance might suggest that anomalous areas of very high uranium content might exist, whereas less variation in the population might make it less likely that the highest values will differ greatly from the mean. The variation between lithological types might be misleading if considered on its own, however. An examination of means and variances for different stratigraphic units might reveal that the variation between the units was considerably greater than that between lithological classes. Perhaps all lithologies in the Ordovician had comparatively high values, and a high mean value and low variance for, say, red shales might reflect the fact that the only red shale samples came from the Ordovician. Grey siltstones, on the other hand, might show high variation simply because they were collected from many different stratigraphic levels. Similarly, stratigraphic age and geographical area are likely to be linked, and the values for one should not be interpreted without reference to the other. The aim of analysis of variance is to identify and measure the various sources of variation. The total variance in a population is split up, or partitioned, into the variance within groups or subsets, such as grey siltstone, grey shale, red shale, striped mudstone; and between groups, that is the variance of the means of the subsets. In order to separate the effects of lithological variation from those of stratigraphic variation, each lithology would ideally be represented in the same proportions in the samples from every stratigraphic class, and vice versa. Since lithology and stratigraphy are two separate ways of dividing the population, an analysis of this kind is known as a two-way analysis of variance. If geographic area is considered as a third independent way of classifying the data, a three-way analysis of variance could be performed, and ideally each lithology and each stratigraphic unit should be represented in the sample in the same proportions in each area. The methods of experimental design which make this possible in a controlled experiment can seldom be applied in geological data collection, and might conflict with other sampling objectives. Visual examination of the table of means and variances by a geologist familiar with the limitations of the data and the collection procedures may therefore be the most fruitful approach to interpretation.

In contrast to the two-way or multi-way presentation of data, the subsets may have a hierarchical structure. Means and variances of uranium content might, for instance, be calculated for the region as a whole, then separately for igneous, metamorphic and sedimentary rocks. The sedimentary rocks might in turn be subdivided into, say, conglomerate, sandstone, siltstone and shale, with statistics calculated for each. An even finer subdivision might be attempted in which, say, shale was subdivided by its colour, or sandstone by its grain size, texture or supposed genesis. The data could be summarized for purposes of interpretation into means and variances of major groups, followed by values arranged by subheadings and sub-subheadings, as in Fig. 56. The method of interpretation would then be to inspect the values to see how they varied among the classes which were part of a group at the next higher level in the hierarchy, and how statistics for each class differed from the group mean and variance. Among other things, this might indicate the degree

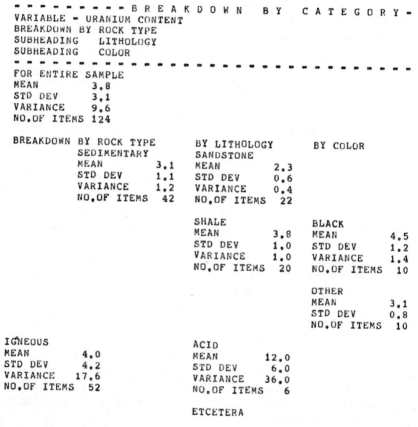

```
- - - - - - - - - B R E A K D O W N   B Y   C A T E G O R Y -
VARIABLE - URANIUM CONTENT
BREAKDOWN BY ROCK TYPE
SUBHEADING     LITHOLOGY
SUBHEADING     COLOR
- - - - - - - - - - - - - - - - - - - - - - - - - - - - - -
FOR ENTIRE SAMPLE
MEAN        3.8
STD DEV     3.1
VARIANCE    9.6
NO.OF ITEMS 124

BREAKDOWN BY ROCK TYPE        BY LITHOLOGY          BY COLOR
          SEDIMENTARY          SANDSTONE
          MEAN         3.1     MEAN         2.3
          STD DEV      1.1     STD DEV      0.6
          VARIANCE     1.2     VARIANCE     0.4
          NO.OF ITEMS   42     NO.OF ITEMS   22

                               SHALE                 BLACK
                               MEAN         3.8      MEAN         4.5
                               STD DEV      1.0      STD DEV      1.2
                               VARIANCE     1.0      VARIANCE     1.4
                               NO.OF ITEMS   20      NO.OF ITEMS   10

                                                     OTHER
                                                     MEAN         3.1
                                                     STD DEV      0.8
                                                     NO.OF ITEMS   10

IGNEOUS                        ACID
MEAN        4.0                MEAN        12.0
STD DEV     4.2                STD DEV      6.0
VARIANCE   17.6                VARIANCE    36.0
NO.OF ITEMS 52                 NO.OF ITEMS   6

                               ETCETERA
```

Fig. 56. Analysis of variance table, giving the mean and variance of categories of the data.

of diversity at each level in the hierarchy, and show which classes had data with unusual characteristics. The hierarchy may be a feature of the classification rather than being expressed explicitly in the data. It might even be determined by cluster analysis after the data were collected. The user may therefore have to provide a complex set of instructions to the computer program to ensure that the data are correctly organized for presentation. It may also be necessary to extend the data set to include new criteria for classification. A rigorous analysis of variance for this kind of data, known as hierarchical analysis of variance, is possible, but again only if the data were collected according to an appropriate experimental design which is seldom practicable in geology.

There are many methods of data analysis which have not been mentioned in this brief review, but the main elements in data interpretation, not all relevant in any one application, are generally: looking for errors and anomalies, and correcting or removing them if necessary; calculation of summary statistics and parameters of the frequency distribution; consideration of correlation between quantitative variables and cross-association between qualitative variables; cluster analysis to look for subsets of similar items, and PCA to look for groups of correlated variables; regression analysis to relate one variable to a set of quantitative variables which may in part account for its properties; or analysis of variance to relate one variable to a set of qualitative variables which may account for its properties. The geologist may be more familiar with pictorial than with numerical interpretation of his data, and diagrams and maps produced by computer on the line printer or graph plotter may be the most effective means of presenting the raw or transformed data or the results of data analysis for visual interpretation. The three-dimensional spatial framework of many geological data sets gives particular importance to the analysis of sequences of data points in space or time, and for this purpose, the techniques mentioned above can be applied in rather specialized ways.

3.17. Sequence and spatial analysis

In the science of statistics, the study of sequences is often approached through time series analysis, which provides methods of analysing a series of quantitative values, such as the annual production of coal over a period of 50 years, or the voltage levels in a line during transmission of a message. In geography, geology and the earth sciences generally, statistical analysis of series in space is extended from one to two or three dimensions, as spatial analysis. Whereas time is normally thought of as moving in one direction only, so that a value can be affected by earlier but not by later events, this does not necessarily apply to sequences in space. Therefore, although time series methods are usually applicable in spatial analysis, they should not be used uncritically.

The spatial data with which a geologist has to deal may be point values, such as measurements at boreholes, or at outcrop; line data, such as geological boundaries or faults; sequences of adjacent values, such as geophysical traverses or measured sections; or orientation data, such as strike and dip measurements of bedding planes. The first step, as usual in computer processing, is normally checking and verification of the data, possibly with programs designed to bring doubtful or unusual values to the user's attention. For example, values that do not plot within the boundary of a map can be printed out, for the user to check their coordinates. The sampling points for stream sediments can be plotted on a transparent overlay and placed over a topographic base map or air photograph to check that they lie on a stream. The elevations of horizons in a borehole can be checked by program to see that they are in the correct order, and above the total depth. A plotted map of seismic shot points can be scanned by eye to see that the sections lie in the expected positions.

3.18. Symbol plotting and value posting; geographic transformation

Display of the data and of the results of the analysis is highly desirable with spatial information, and although plots can be produced on a line printer, a graph plotter gives much better results. Point data, such as the location of boreholes, or elevations of a horizon at a set of boreholes, can be plotted on a map as symbols, with or without a name or value written alongside each one, see Fig. 57. The symbol may simply indicate the location of the point, or different symbols can be used to distinguish items such as gas wells, oil wells and dry holes. Symbols with quantitative significance can also be plotted (see Fig. 58), with the size of the symbol indicating the percentage of a chemical element, or the orientation of a strike and dip symbol representing a structural measurement (Fig. 59). Values posted at the correct locations on a set of maps, each of which shows, for instance, the elevations of one stratigraphic horizon, can save time, as a preliminary to hand contouring. Programs of this kind usually require the user to select the scale, area and title of the map, the required symbols and their size, the size and orientation of the captions, and to decide whether or not grid lines should be shown and labelled. Where the user does not make a choice, a standard 'default' option may be invoked by the program.

The plotting subroutine is likely to have been written to accept the locations as x and y values giving the distance to the right and above an origin at the bottom left-hand corner of the plot. It is likely to have facilities for calculating these values from eastings and northings. The ability to make

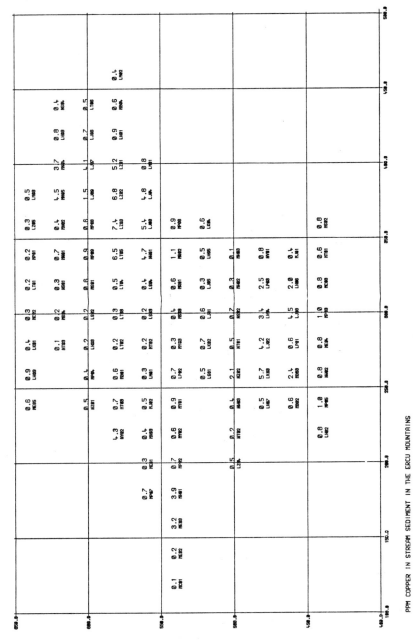

Fig. 57. Diagrammatic map of point data with posted values.

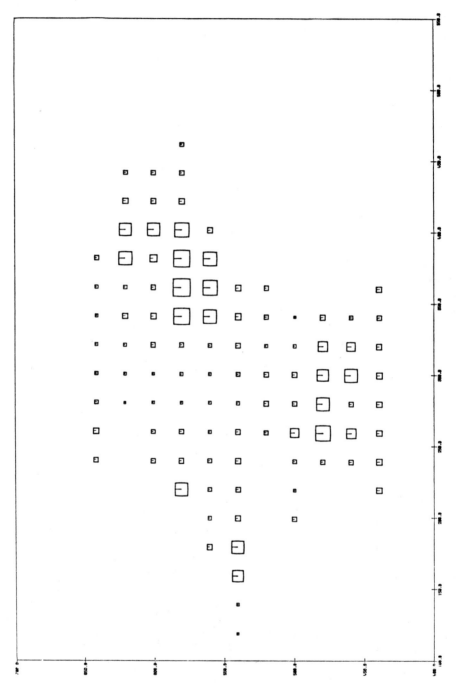

PPM COPPER IN STREAM SEDIMENT IN THE ERCU MOUNTAINS

Fig. 58. Diagrammatic maps in which the area of the symbol represents the value at that point

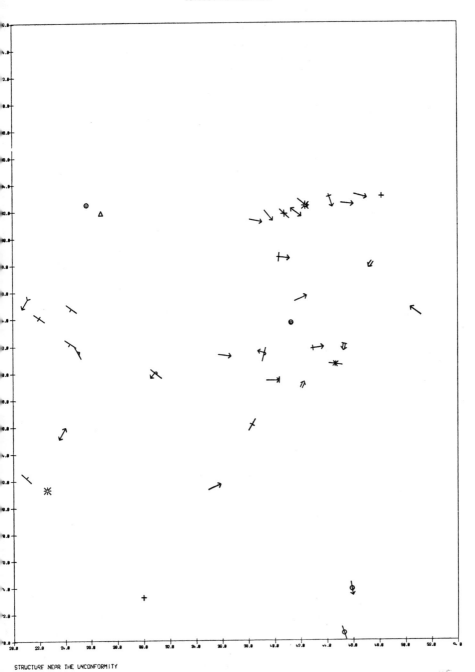

STRUCTURE NEAR THE UNCONFORMITY

Fig. 59. Diagrammatic map of orientation data.

geographical transformation to this form from other reference systems may be built in to the routine. If not, separate programs are available for handling the transformation, for example, from LSD, township and range (Good, 1964), from satellite and other navigational systems, and from other map projections, such as Universal Transverse Mercator. Other, more complex, transformations are possible by cross-linkage between files. For instance, given access to a file with the appropriate information, location in space could be transformed to hydrological position expressed as proportional distance upstream from a tributary junction of a particular stream order. This could have application to the analysis of geochemical stream sediment samples, as well as to hydrology and hydrochemistry.

3.19. Moving averages; grey-scale maps; smoothing and filtering

From scattered point data displayed as a map, it is a natural step in spatial analysis to try to detect and display pattern in space. The use of trend surface analysis has been mentioned, and automated contouring is considered later in the context of explanation. These techniques are appropriate for more or less continuous surfaces which are comparatively smooth over short distances, so that two points that are close to one another are likely to have more similar values than two points further apart. But for some data, such as geochemical analyses of stream sediments, such methods may be inappropriate. Although the distribution of elements in stream sediments presumably reflects pattern in the underlying geology, the population of possible samples is discontinuous, with a real existence only in the stream beds and with considerable local variation related to such aspects as position across the stream. A smooth regional surface of the kind depicted in a conventional contour map therefore does not exist. Nevertheless, a method, known as moving average, has been found to be useful in smoothing the small-scale local variation and revealing the overall pattern more clearly (see Davis, 1973).

The procedure can be visualized in terms of a map on which the data values are posted. Conceptually, an opaque overlay is placed on the map, with a small round window sufficiently large to reveal a small number of data points, see Fig. 60. The window is placed initially at the top left-hand corner of the map, and the average value of the data points visible through the window is calculated and taken as the moving average value for the center of the window. The window is moved to the right by a distance equal to a specified fraction, perhaps one third, of its width, and a new moving average is calculated and taken as the value for the center of the window in its new position. The procedure is repeated step by step across the map, until the right-hand side is reached, then the window is moved down by the step

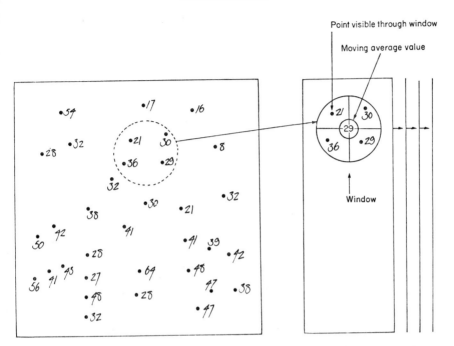

Fig. 60. Calculation of moving average. The window, shown on the right, is scanned step by step over the map shown on the left. At each step, the values visible in the window are averaged, and the average value plotted to give a grid of 'moving average' values.

distance, and the process repeated for the next row, and so on until the entire map is scanned and a complete grid of moving average values is calculated. The process can be carried out mechanically, by moving a card with a circular hole systematically across the map, but is less laborious and more accurate by computer. The moving averages can then be plotted as a set of grid values. Alternatively, they can be displayed as a grey-scale map (see Fig. 61) in which the values are represented by shades of grey produced by overprinted characters on a line printer or patterns of dots produced by a raster plotter.

The moving average method and grey-scale display can be applied to any set of values showing a regional pattern obscured by local variation. For example, a moving average map could represent the pattern of sampling density. The technique, by smoothing over discontinuities, generates a continuous pattern from a discontinuous one. This applies to the earlier examples, but many other types of variable can be measured through a scanning window and plotted as a moving average map. Examples are the percentage area of outcrop, the length of fault line, the average throw of faults, the orientation of bedding planes or formation boundaries, the curvature of

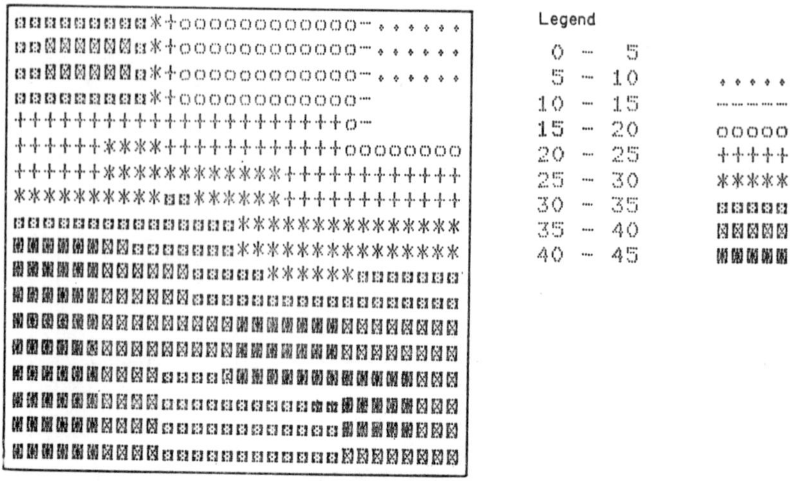

Fig. 61. A grey-scale map, in which higher values are represented by darker shades of grey.

streams or their frequency of branching. The advantages of representing them as continuous variables are, first, that their pattern in space may be more easily visualized; second, that they can be compared directly with continuous surfaces, such as contours on the basement or a hypothetical stress pattern; and third, that they can be analyzed and displayed further by methods of spatial analysis designed for continuous surfaces. The procedure of moving averages is one of a family of techniques for smoothing, that is, removing the small-scale local variation, with the objective of revealing larger scale pattern more clearly.

In communications theory, methods of separating and emphasizing components of a series which carry useful information (the signal) as opposed to other components which obscure the signal (the noise), have been studied extensively because of their importance in such fields as telephone and radio transmission. In addition to the smoothing procedure, in which the noise is assumed to be in the high-frequency components only, various methods of filtering have been developed, in which a specified range of frequencies is emphasized with frequencies outside the range either damped or removed. A filter of this kind operates when a radio is tuned to receive a particular station. The analogous case with geological pattern is the selection and isolation of components of variation on different scales to examine the geological significance of each. The technique of trend surface analysis is an example of a filter which separates large-scale variation or trend from small-scale variation, or residual, where both trend and residual may have meaning for the geologist.

3.20. Transition probabilities

Much of the sequential information in geology is qualitative, such as measured outcrop sections, lithological or stratigraphical sequences from boreholes, or even the pattern of formations shown on a conventional geological map. For one-dimensional series, or for sequences from several boreholes plotted side by side, computer display (see Fig. 62) can be helpful. The pattern of change in a qualitative series can be described in terms of the relative frequencies of the different changes or transitions. Thus, if

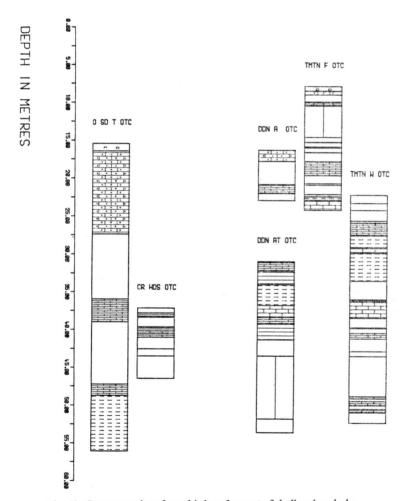

Fig. 62. Computer plot of graphic logs for a set of shallow boreholes.

L

Bed above						
Bed	Conglom-erate	Sandstone	Siltstone	Shale	Limestone	Total
Conglomerate	0	5	3	2	0	10
Sandstone	0	0	1	18	1	20
Siltstone	1	1	0	16	2	20
Shale	8	14	11	0	7	40
Limestone	1	0	5	4	0	10
Totals	10	20	20	40	10	100

Fig. 63. Table of transition frequencies, comparable to the contingency table of Fig. 55.

sandstone is overlain by shale at 18 levels in a measured section, then the frequency of sandstone to shale transition is 18. The transition frequencies for all the lithologies in the section can be recorded on a transition table or transition matrix. The headings on the rows of the transition table of Fig. 63 are the lithologies of the various beds in the sequence. The headings on the columns are the lithologies of the overlying beds. It may be noted that the transition table is a special case of the contingency table (Fig. 55), in which the characteristics being compared are, for example, the lithology of each bed and of the bed above. For computational purposes, the same routine can be used if, as a preliminary step, the data, arranged as a column of an array, are repeated as a second column, each bed one row lower than before. The contingency table computation, which compares two sets of attributes of an array, is then comparing the attributes of each bed with the overlying bed. As with contingency tables, there may be a need to compare transitions as relative rather than absolute frequencies (see Davis, 1973). With 101 beds and 100 transitions in the section, the relative transition frequency for sandstone to shale is 18/100 or 0·18. This still reflects the relative abundance of sandstone and shale, as opposed to other lithologies, in the section, and if the point of interest is whether the tendency for a shale bed to occur over a sandstone bed is greater or less than say, over a limestone bed, then a comparison can be made with the values expected if the lithologies occurred independently (see Fig. 55).

Several attributes may be measured for each bed in a succession. They may be qualitative data, such as the presence or absence of each of a number of fossil species; semi-quantitative, such as the relative abundance of the species; or quantitative, such as a set of well logs. Again this information can be displayed graphically, see Fig. 64, and contingency and correlation

SWNSWCK 15

DEPTH IN METRES

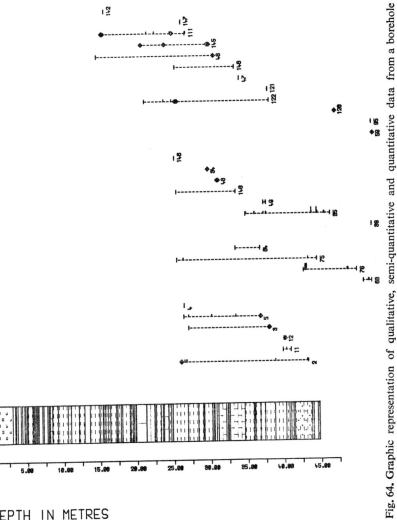

Fig. 64. Graphic representation of qualitative, semi-quantitative and quantitative data from a borehole

tables can be computed to show the interrelationships of the attributes or variables. If the data are quantitative rather than qualitative, the correlation coefficient matrix replaces the transition probability table. In long sequences, transition frequencies can be calculated not only for the following item, such as the overlying bed, but also the next following item, the one after that, and so on. If several items in succession are considered, it is again possible to use standard procedures for computing contingencies, correlations, or even multiple regressions, by repeating the column of data values as often as required. setting the column one step lower each time (see Fig. 66).

3.21. Serial correlation; correlograms

Smoothing and curve- or surface-fitting techniques consider data points within the frame of reference of the series or surface of which they are a sample. An alternative approach is to develop descriptors of the surface considered as one of a family of possible surfaces. The descriptors must be based on the relationships of points on the surface to one another, as this is the only relevant information available, and should describe geologically useful aspects of the surface. If this can be achieved, then geological surfaces of different types, or of the same type from different depths or areas, could be compared and contrasted. Furthermore, information from the larger frame of reference, namely knowledge of the properties of a wide range of geological surfaces, could be brought to bear on the estimation of properties of a particular surface, such as the volume beneath it. A number of statistical methods have been used for describing the relationship of points on a surface.

One starting point is to calculate the autocorrelation or serial correlation, which is the correlation coefficient of data values between points on a sequence or surface at a specified distance apart (see Davis, 1973). Serial correlation is a function of distance, and the manner in which it varies with distance is a reflection of the form of the surface. The serial correlation function can be plotted as a graph, known as a correlogram, of correlation coefficient against separation distance. This is most easily calculated if data values are available on a regularly spaced grid, otherwise the number of point pairs at some separations may be too small to give a reasonable estimate of the correlation coefficient. A high value of the correlogram at a particular separation distance indicates that the values of two points at that distance apart are likely to be similar, or at least more similar than points with separation distances that have a lower value on the correlogram. In geometrical terms, the similarity would indicate that the points are at about the same height on the surface. In the examples considered earlier, the smoothly folded stratigraphic horizon would have high values of the correlogram over short distances, decreasing as the influence of the large folds is

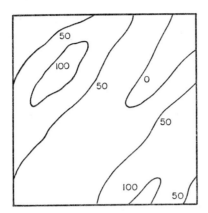

Fig. 65. Contour map of a surface with a strong grain in a northeasterly orientation.

felt. The correlogram for geochemical data from stream samples would decrease sharply at short separation distances, reflecting the sharp local variation within the stream. Thus the correlogram to some extent characterizes properties of geological significance.

Many geological and topographical surfaces have a grain (see Fig. 65) which implies that variation is more rapid on traverses in one orientation than in others. The shape of the correlogram thus depends on the orientation in which separation distance is measured, and its changes with orientation are another reflection of the form of the surface. Correlograms for surfaces are thus generally computed for at least two orientations, generally at right angles to one another. The correlogram is equally applicable in one dimension, for example for well logs, and can also be used in three dimensions, for instance to depict the variation of grade in an ore body of considerable vertical as well as horizontal extent. Three separate correlograms represent each of the three dimensions. Because the correlograms are related to distance, the properties of surfaces from different areas can be compared by comparison of the correlograms. The correlogram is however very sensitive to sampling errors, is particularly difficult to construct with irregularly spaced data, and interpretation can be difficult. Any trend, such as a regional slope on a surface, should be removed before the correlograms are prepared and compared. The distinction between trend and residual in trend-surface analysis, however, is not easy to make. Furthermore, the orientation of the correlograms is somewhat arbitrary, probably depending on the choice of sampling pattern. The methods of 'kriging', developed by French workers, use information of this kind in preparing estimates of the volume and variability of ore reserves (see Harbaugh et al., 1977). Because of the difficulty of arriving at accurate estimates of the relationships from a small sample, however, the results when applied to surface analysis may be no more accurate than the geologist's estimate based on his background geological knowledge.

In one-dimensional series, such as well logs or geophysical traverses, the frequency of sampling may be very high, and high-order power series or trigonometric series can be fitted.

Vertical series in particular may show a cyclicity that can be suitably represented by a trigonometric series. Such a series can be regarded as a set of terms each defining a regular sine wave of different amplitude (height) and wavelength, which, when added together, give a curve similar to the original data. The relative importance of the various wavelengths may be apparent from a correlogram, and if appropriate, can be studied further by spectral analysis. The original curve can be reconstituted with a certain wavelength, or range of wavelengths, filtered out. Although spectral analysis and filtering have a place in geophysics, repetition of events in geology seldom has the precise periodic nature of sine waves with a superimposed random component. More often, the random events are integrated into the

3·47	↓		
43·73	3·47	↓	
86·36	43·73	3·47	↓
96·47	86·36	43·73	3·47
36·61	96·47	86·36	43·73
.	.	.	.
.	.	.	.
.	.	.	.
46·98	.	.	.
63·71	46·98	.	.
62·33	63·71	46·98	.
71·62	62·33	63·71	46·98
33·26	71·62	62·33	63·71
	33·26	71·62	62·33
		33·26	71·62
			33·26

Fig. 66. The same string of data is set one row lower in each successive column. Auto-regression and autocorrelation can be examined in the matrix of complete rows, using standard regression or correlation subroutines. The incomplete rows at each end cannot be included, causing errors known as edge effects.

sequence so that the periodic components are out of phase in different parts of the sequence. Fitting a trigonometric series is then inappropriate. Instead, the autoregressive method may be relevant, in which a regression equation relates each value to the preceding set of values in the series. The distinction is clearly presented in a statistical context by Yule and Kendall (1958). The computation can be visualized as multiple regression in which the matrix of independent variables is constructed from the column of data repeated several times, and set one step lower each time (see Fig. 66). The data sets at either end of the sequence are incomplete and so cannot be used in the computation.

3.22. Slope analysis

The frequency distribution of slopes on a surface may be of interest for a number of reasons. A knowledge of slopes on the present-day land surface may be relevant to applications in engineering geology for predicting landslips, in ecology and agriculture for predicting intensity and duration of sunshine on the surface, in highway engineering for selecting a route for a new road, in geomorphology for linking to rates of erosion, and to the field geologist for identifying features that could reflect the underlying geology. The structural geologist may be concerned with many layers of folded strata, rather than a single surface, and his raw data are likely to be in the form of orientation measurements. Even if only a single surface is considered, however, the areal distribution of slopes may indicate where faults and other discontinuities are likely to be present. An initial step in some contouring programs is the calculation from elevation data of average slopes over small areas, usually by fitting a local linear trend surface (see Harbaugh et al., 1977). The orientation of such slopes can be represented by the direction cosines of a line normal to the surface. The orientations can be represented on a stereogram (see Koch and Link, 1970) and, can be analyzed by methods such as those outlined in considering geometric data processing (§3.13) and the analysis of orientation data (§3.14). Apart from the direct applications in slope analysis, future developments in this field seem to offer a means of linking information about elevations of a surface from boreholes and outcrops, with the structural geologist's orientation data. Slope analysis may also prove to be of future interest as an alternative to autocorrelation analysis, for developing descriptive parameters for the properties of surfaces.

In summary, statistics, such as regression and correlation methods, can be adapted to analyze sequences in time or space, in one, two or three dimensions. A geological surface, from this point of view, is a sequence in two dimensions. Instead of considering several variables measured at each of a number of points, spatial analysis is concerned with the relationships among adjacent areas, as in moving average calculations, and among adjacent points, as in correlograms. Smoothing can be performed to remove small-scale variation and reveal larger trends more clearly, or components on different scales can be examined separately by filtering. The most successful and widely used method of spatial analysis is trend-surface analysis in which a trend, or surface showing only large-scale regional variation, is considered separately from the local residuals. For this purpose, regression methods are used to find an overall fit of a two-dimensional power series to the data points. Other methods of surface analysis consider the relationship of data points to one another, and thus provide a means of comparing the properties of different sequences and surfaces. The relationships can be described by a transition matrix for qualitative data, and a correlogram for quantitative data, with the stereogram a possible extension to the third dimension.

4. EXPLANATION

4.1 The nature of explanation; the place of the computer

Explanation can take many forms, but is generally concerned with placing phenomena or observations in a broader framework so that their meaning and significance within an established set of postulates and theories is clarified. For instance, a sample of observations of grain size of modern beach sediments might be 'explained' by relating them to an energy pattern in the depositional environment. Interpretative statistics from geochemical data for a set of stream sediments might be 'explained' in terms of bedrock geology and its mineralization, the drainage pattern, the disintegration of the solid rock and its dispersal as sediment, the solubility and removal of certain constituents, and the sampling scheme and methods of chemical analysis which produced the data. Geological explanation frequently takes the form of a description of a sequence of historical events taking place in a particular setting, which could account for the data that have been collected and observations that have been made. The explanation, in the spirit of Hutton's Principle of Uniformitarianism, would be expected to be consistent with the behavior of geological processes as far as they can be ascertained from present-day observations and experiments, and with the postulates of physics, chemistry and biology. It would also be expected to be consistent with any relevant larger-scale geological explanations. For example, an account of sedimentation in an area believed to have been on the margin of a continental plate, would be expected to be consistent with the anticipated behaviour of a depositional area in that position. Any inconsistency might be resolved by amending the account of the local phenomena, or by reconsidering the larger-scale explanation. The real world, and the observations and data obtained from it, on the other hand, are not, by the conventions of science, altered to fit an explanation. Observations which do not agree with an explanation may of course be mistaken and data may be inaccurate, and, if this is established, the data are either discarded or corrected. The test against the evidence of the real world, nevertheless, remains the final arbiter of the acceptability of an explanation.

The historical account, which is the form taken by many geological

175

explanations, makes heavy use of the concept of cause and effect. An example might be: 'Because the mountains in the east were uplifted more rapidly than those in the west, there was more vigorous erosion of the eastern mountains, and coarser sediments were deposited at their base.' Or: 'Because there was a higher proportion of mud relative to sand in the sediments at the west end of the bay, the amount of compaction was greater, and peat beds there accumulated to greater thicknesses during periods of constant sea level.' Because of a set of causes, a particular effect followed, thus creating the conditions for the next geological event. The consequences of such events which can be observed in the rocks represent effects rather than causes, and the latter are known only by inference. In most real geological examples of any complexity, the set of causes cannot be determined uniquely from observing the effects. Although similar causes may produce similar effects, the reverse is not true, and the same effect can be produced by many different combinations of causes. To take a simple mathematical analogy, the computer might generate the number $6 \cdot 0$ as a result of an arithmetical operation on two numbers. Without further evidence, there is no means of knowing whether the operation was $2 \cdot 0 \times 3 \cdot 0$ or $2 \cdot 0 + 4 \cdot 0$ or $0 \cdot 0 + 6 \cdot 0$. Even the advice to take the simplest explanation as a working hypothesis does not help. On the other hand, if the set of causes is known, the outcome can be predicted with greater certainty. If the computer was known to have added $2 \cdot 0$ and $4 \cdot 0$ correctly, the result would be a foregone conclusion. The geologist cannot arrive directly at an explanation, in terms of hypo- thetical causes, by interpretation or analysis which merely transforms the data into a more compact or a more comprehensible form. The process of explanation is one of imaginative inductive thinking where instinct, pre- conceptions, analogy, intuition and even aesthetic judgement all have a part to play. It is an activity in which the human mind stands supreme, and to which the computers and software at present used in geological applications can make no contribution. The computer can, however, interpret and display the information to assist the geologist in formulating his explanations. The computer can also help him to explore the consequences of his hypo- theses and explanations, and it is with this latter activity that the present chapter is concerned.

Perhaps the first reason that can be given for considering at some length the place of the computer in geological explanation is that at present the various subsystems of geological investigation listed in §1.5 are, from the systems viewpoint, unbalanced. Modern methods of geochemical analysis, geophysical logging, remote sensing and the like have greatly increased the amount of quantitative data available to the geologist. Even from conven- tional sources, such as field mapping, the amount of data collected often exceeds the ability of the geologist to assimilate it. Computer methods of data interpretation have helped to rectify this imbalance, enabling the geologist to extract more of the useful information concealed in his data. There seems to be good reason to suppose that his trend will continue, as processing systems become better, cheaper, more accessible, and easier to use, and as geologists become more familiar with the techniques and more adept at collecting data by methods which ensure that they can be fully analyzed.

The geological explanation of the data so laboriously collected and analyzed is at an entirely different level. In contrast to the quantitative precision of the data, the explanation may consist of no more than broad qualitative statements. Many a research thesis, with the objective of elucidating events in the geological past, has reached conclusions of the type: 'On the basis of five hundred detailed analyses of size distribution in sediment samples, it was concluded that the grain size decreased from west to east, and that the source of sediment may therefore have lain in a westward direction.' 'The ten thousand strike and dip measurements, precise to 2°, were plotted on a stereogram and suggest that the main axis of folding had a northerly orientation.' The data are far more precise and detailed than the explanation warrants. Computer methods give promise of being able to define an explanation, or range of explanations, and explore their implications in precise quantitative terms which more nearly match the detail of data collection and interpretation.

A second reason for attempting greater quantification in the explanatory stage is the sheer complexity of geological processes, and the resulting likelihood of being misled by simplistic explanations of the results of data interpretation. A multimodal frequency distribution of the abundance of geochemical elements in stream sediments, for instance, might arise from aspects of dispersion in streams or of sampling schemes, and unless the likely effect of these factors can be estimated and allowed for, it would be unwise to base an explanation solely on postulated variations in the bedrock. A third reason for considering that additional effort in this area is justified is its success in other subject areas, such as ecology and economics, where analogous problems are encountered. The use of computer models has made it possible to explain the behavior of natural systems in a more complete, precise and detailed manner than before, and to predict more accurately the response of the system to external changes or to its own internal development. On the other hand, much of geology remains a descriptive science, in which data are collected without the benefit of any widely agreed model, in the expectation that explanations will be amended and modified during subsequent investigations. This may reflect the difficulties of establishing an exact knowledge of relationships and processes in areas where experimentation can be difficult or impossible. The computer models that can provide a framework for collecting, interpreting and explaining data in subjects like geophysics, ecology or meteorology have not yet been developed in the more traditional areas of geology. Although the techniques are well established, therefore, it is not always possible to point to their successful application in geology, where their use should perhaps still be regarded with the caution appropriate to an area of current research rather than routine application.

There are perhaps three main aspects of explanation in which computer methods are involved. First, the results of data analysis can in some cases be tested against mathematical deductions. Descriptive statistics derived from a sample, for example, may bear a known relationship to those of the population, and if certain assumptions can be made, then deductions from probability theory can be tested against statistics derived from the data. In other words, deductions from a quantitative explanatory model can be tested against observation. A second aspect is the use of simulation methods to explore the consequences of several possible hypotheses, where they cannot readily be determined by mathematical deduction. Third, systems methods offer a means of organizing a complex set of interrelated hypotheses within a framework where the component parts and their relationships can be recognized and evaluated.

4.2. Probability, simulation, uncertainty and risk

The first of these aspects depends on probability theory, which is the basis of that part of the science of statistics that is concerned with analyzing the likelihood of occurrence of certain events and combinations of events, and with the description in algebraic form of frequency distributions of variables with an unpredictable or random component. Statistics considers how the properties of a sample drawn from a known population can best be described, and hence, how, from a sample of measurements, conclusions can be drawn about the properties of the sample, and the likelihood of its having been drawn from a particular population. It might be expected that this form of reasoning would be of major importance to geologists. In fact, the statistical applications that have so far proved to be most important in geology are the methods of data analysis described in the last chapter. The reason may be that deductions in terms of probabilities depend on assumptions about the frequency distributions and the nature of the relationships between variables in the population from which the sample is drawn. These are not necessarily valid in geological investigations. Even the theory of sampling may not be helpful if the geologist can only collect data where conditions allow. The difficulty of conducting experiments relevant to the scale, duration and conditions of geological processes means that the geologist can seldom control the variables that he measures, but can merely record what he can observe, probably many millions of years after the process is complete. Nevertheless, the concepts of probability and of calculating precise quantitative measures of the properties of inexact data are widely applicable in geology, and many of the measures are sufficiently robust that deductions can be drawn from them, even if the basic assumptions are not entirely met.

The basic assumptions in much statistical reasoning concern the frequency distributions of the variates, and their relationships. The mathematical background, expressed in probability theory, is described at various levels of detail in the early chapters of many statistical textbooks as well as in works dealing specifically with the subject. A brief review of some of the ideas is given here, as they underlie the methods described in this chapter. A starting point in considering probabilities is the familiar activity of tossing a coin. The complex interplay of influences which are unknown and maybe unknowable, but evenly balanced, results in an equal chance of the coin landing 'heads' or 'tails'. Similarly, throwing a dice gives an even chance of each of six possible outcomes. The probability of one specific event, such as obtaining a 3 on one throw of the dice, can be measured on a probability scale from $0 \cdot 0$, meaning that there is no chance of the event occurring, to $1 \cdot 0$, meaning that the event will certainly occur. Since there are two equally probable outcomes of tossing a coin, each has the probability $0 \cdot 5$, and each of the six possible outcomes of throwing a dice has the probability $\frac{1}{6}$. The algebra of probabilities is based on two procedures for manipulating probabilities, namely addition and multiplication. The probability of occurrence of one of a number of events is equal to the sum of the probabilities of the individual events. Thus the probability of a tossed coin coming down either heads or tails is $0 \cdot 5 + 0 \cdot 5$ or $1 \cdot 0$ (absolutely certain). The probability of a thrown dice showing either 1, 2 or 3 spots is $\frac{1}{6} + \frac{1}{6} + \frac{1}{6}$ or $\frac{1}{2}$. The other procedure is the multiplication of probabilities to determine the probability of several independent events all occurring. The probability of tossing a coin and obtaining a head on the first throw and a tail on the second is $\frac{1}{2} \times \frac{1}{2}$ or $\frac{1}{4}$. The probability of obtaining six 6's in succession on throwing dice is $\frac{1}{6} \times \frac{1}{6} \times \frac{1}{6} \times \frac{1}{6} \times \frac{1}{6} \times \frac{1}{6}$ or $1/46656$. The laws of addition and multiplication of probabilities assume that the individual events are independent and unaffected by previous events. If six 6's had been thrown in succession, the theoretician would still regard the probability of the next throw being 6 as $\frac{1}{6}$. The practical man would suspect loaded dice. The unlikely events of the coin landing on edge, or the dice failing to land at all, are ignored as being irrelevant. Where genuine ambiguity could exist, however, it is advisable to define precisely the set of outcomes under consideration.

The coin and the dice are familiar enough to help in visualizing the rather abstract concept of probability. They also provide a means of experimenting with hypotheses involving probabilities. The probability of tossing a coin twice and obtaining a head the first time and a tail the second is $\frac{1}{4}$. But this does not indicate what the outcome will be if the experiment is performed. It is because the result of any one trial cannot be known in advance that probabilities, and not certainties, are involved. If, in an experiment, the head and tail sequence was obtained, there would be no reason for surprise. There was one chance in four that this would happen. If it did not happen, there would be even less reason for surprise, since the probability was $\frac{3}{4}$. Nevertheless, although the precise outcome of these events cannot be predicted exactly, the probability statement does describe the long term behavior of repeated trials of the experiment. If the sequence of two tosses of the coin was repeated many times, the average number of head–tail combinations would be close to $\frac{1}{4}$ of the total. The more often the experiment was repeated, the more confident one could be of the average ratio being nearly $\frac{1}{4}$. Probabilities are thus measures of a tendency that is expected to be displayed over a large number of events, even though the outcome of an individual event is uncertain. This principle of the 'central tendency', of the predictability of the characteristics of a large number of uncertain events, is the basis for the economic survival of insurance companies and bookmakers. The loss of an individual car, ship, life or horse race is unpredictable, but the statistical results of a large number of events can be predicted with considerable accuracy, and the unpredictable risk to an individual becomes predictable in aggregate to the company. It is important to note, however, that actuarial calculations are based on assumptions about the continuation of an existing set of conditions. An outbreak of war, for example, could alter the assumptions on which an insurance firm calculated probabilities, as is recognized implicitly in the small print of many an insurance policy.

The idea of probability can be extended to events which are not repeatable. What is the probability of a particular glass breaking if it falls over? What is the probability of encountering Jurassic strata by drilling on a particular location? Both questions require a prediction to be made on the basis of uncertain knowledge, but are concerned not with repeated trials but with a single event. The uncertainty, and the probability, thus refer to a conceptual model, rather than to the real event. Within that model, one can suppose that there are unpredictable elements, like the air currents around the glass as it falls and stress patterns within the glass related to its manufacture, age and temperature, that would determine whether the glass breaks. Similarly, unpredictable elements in geological events during deposition, erosion and structural deformation might determine whether Jurassic strata were present or not. These irregular or random elements make it impossible to answer the questions exactly. However, by analogy with the dice, one might suppose that there were elements of regularity and pattern in the process which would be a stable element if the process, or model, were repeated many times. The statement of probability in these circumstances is reflecting a degree of belief in a particular outcome on the basis of available knowledge. It is a reasonably familiar intuitive concept, since an everyday activity like driving a car involves frequent assessments of risks and likelihood of occurrence of various events, from a mechanical failure to a child running into the road. The process of simulation, described later in this chapter, provides a means of expressing a hypothesis about a process, such as deposition of Jurassic strata, in quantitative terms which embody a random, unpredictable element, and make it possible to perform a mathematical experiment which has an uncertain outcome reflecting the random element in the model. The results of repeated simulation might show a regularity in the outcomes over many runs, which could be measured as the probability of that outcome. It is thus possible to reconcile intuitive concepts of probability with the mathematical ideas of relative frequency.

These ideas can perhaps be made clearer by describing a simple simulation of a geological event. Considering the problem of estimating the probability of encountering Jurassic strata at a particular drilling location, a simple model might be envisaged which took three factors into account. First, the probable environment of deposition, as reflected in the facies distribution in the area, might give an indication of the lithology. Second, the pattern of subsidence might be a guide to the thickness of strata deposited. Third, the pattern of erosion, related to lithology, might give an indication of whether the Jurassic rocks were likely to be preserved. The purpose of the model is to illustrate a methodology rather than to consider geological events which are specific to the problem. Suppose that examination of the surrounding borehole data indicated that two types of lithology were equally likely to be present in the Jurassic, say a sandstone and shale facies; that study of isopach maps, corrected for the effect of later erosion, indicated that the original thickness of sediment at the location was probably 120 feet, and almost certainly between 80 and 150 feet; and that 100 feet of strata would almost certainly have been eroded at that point, while, if one of the river channels related to the Cretaceous unconformity crossed the site, an event with a probability of 1 in 3 if the Jurassic is of shale facies, 1 in 5 if of sandstone facies, then the Jurassic would be completely removed.

Without recourse to the computer, the system can be simulated mechanically. The outcome of any one trial or simulation, which takes into account the various uncertainties, is itself unknown. Nevertheless, after many trials, a long-term regularity or pattern could be expected. Again for the sake of illustration rather than of recommending a procedure, it is of some interest to consider a method of mechanical simulation of this model. The equal probability of sandstone and shale facies could be simulated by tossing a coin, with, say, heads representing sandstone, tails representing shale. The determination of the thickness of sediment is more difficult to simulate. Any thickness between 80 and 150 feet is regarded as possible, with 120 feet the most likely value. An approximate probability curve could be drawn on this basis, as shown in Fig. 67. One possible mechanism to simulate this would be to obtain counters of equal size and of seven different colours, each colour representing a different ten-foot interval. Estimating the relative frequencies of each interval from Fig. 67, the counters could be selected in the appropriate proportions, thoroughly shaken in a bag, and a counter pulled at random to determine the thickness for that trial.

An alternative approach to simulating events of unequal probability would be to make a transformation of the probability distribution. Within the limits of accuracy of the model, visual inspection of the curve might suggest a subdivision into ten thickness intervals of unequal size, each of which includes an approximately equal area below the probability curve. A method can then be readily found for selecting one of the ten intervals with equal probability, such as throwing two dice, subtracting one from the result, and ignoring a double six. The midpoint of the thickness interval, or more accurately, the value bisecting the area, would be an adequate approximation for this method. Having determined the lithology and the thickness before erosion for one trial, the next task is to consider the likelihood of erosion. As the area was all eroded to a depth of 100 feet, if less than that thickness of strata was deposited, none is expected to remain. But if there was more than 100 feet, the probable outcome depends on the presence or absence of a river valley, and this in turn depends on the lithology. This is known as conditional probability, where the probability of one event depends on the outcome of a preceding event. The choice is shown as a flow diagram in

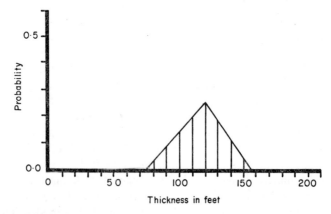

Fig. 67. Estimated probabilities of thickness of sediment. The most likely value is thought to be 120 feet, and the thickness is believed to be not less than 80 feet, and not more than 150 feet.

Fig. 68. If the earlier part of the simulation had selected shale, the ⅓ probability of a river channel removing the Jurassic could be simulated by throwing a dice and arbitrarily considering a 1 or 2 to indicate removal. If the trial selected sandstone with a ⅒ probability of removal, then a 1 could be taken to represent erosion and a 6 regarded as an invalid number, to be repeated until a valid number is thrown.

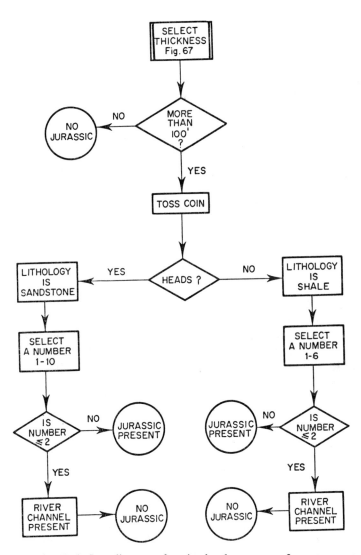

Fig. 68. A flow diagram of a simulated sequence of events.

M

On repeating the trial many times, a definite pattern should emerge in the outcomes, with a stable ratio developing between outcomes indicating the presence or absence of Jurassic. This ratio could then be taken as the estimated probability of encountering the Jurassic at that location. An attempt to perform a mechanical simulation as described will quickly indicate the laborious nature of the task, particularly since many trials are needed to obtain a stable result. It cannot therefore be recommended as a practical procedure. Nevertheless, the same principles can be used in a computer simulation, which is much easier to perform repeatedly, and is much more flexible in both the type of model and the variety of the results. Before considering further the value of simulation methods in geological explanation, however, the important topic of deductions based on probability theory must be considered.

4.3. The normal distribution model

A theoretical frequency distribution of fundamental importance in statistics is the so-called normal distribution, which is followed approximately by many naturally occurring variates. A frequency distribution plot of, say, the lengths of a large number of brachiopod valves in a representative sample of one species, might display the approximate form of the normal distribution (see Figs. 35 and 19). The plot is a symmetrical bell-shaped curve. The mean or average value coincides with the centre of the distribution, known as the median, and with the most frequently occurring value, known as the mode. Values are less frequent the further they deviate from the mean. Some of the properties of the normal distribution were mentioned in §3.8. It is taken as a model in many statistical explanations, and it is therefore relevant to see how geological processes might give rise to such a distribution. Perhaps the simplest way to visualize the development of a normal curve is in terms of a number of small random influences which deflect the value of the variate from its expected value. The expected length of a brachipod valve, for instance, might be 50 mm, but numerous, evenly balanced, random influences throughout the growth of the shell might cause it to grow less or more than would normally be the case. The model can be visualized in mechanical terms as shown in Fig. 69, where a ball enters the apparatus at the top, and would fall directly into the central bin, except that it strikes a peg. Each peg deflects the ball with an equal probability ($0 \cdot 5$) of deflecting it to the right or the left. If many balls are passed through the system, the final pattern of distribution among the bins approximates to the normal distribution. With a very large number of bins and pegs to model the numerous random influences on the development of brachiopod valves, a close similarity might be obtained to the distribution of the continuous variate of valve length.

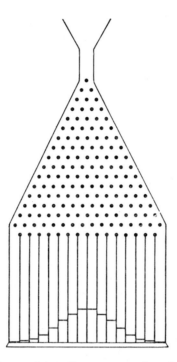

Fig. 69. Granular material poured into the apparatus through the funnel at the top is deflected randomly by a series of pegs before accumulating in the bins at the base. The pattern in the bins resembles a histogram of the normal distribution.

The result does not depend on each peg dividing the sample on an exact $0 \cdot 5$ probability, as long as the overall probabilities are evenly and symmetrically distributed.

4.4. Statistical inference; tests of significance and the null hypothesis

This method of generating a normal distribution suggests that its properties can be precisely defined in probability terms. This is indeed the case, although the mathematical description of the curve, and its derivation, are quite complex. It is therefore possible to predict, for a variate with normal distribution and known parameters, the probability of a measurement lying between defined limits, represented by the area below the curve between these limits. For instance, 68% of the measurements lie within one standard deviation of the mean in a normal distribution, 96% within two standard deviations, and $99 \cdot 7\%$ within three. Probability theory is based on ideal cases which can be handled mathematically, and whose properties are known. In practical applications, the reason for measuring the variate is to estimate its properties, which are certainly not known in advance. Statistical statements therefore tend to refer to an assumption, rather than directly to a sample. If the assumptions are reasonable, then deductions can be made about the expected properties of a sample drawn from the hypothetical population, and these can be compared with the properties of a sample of actual measurements.

As an example, the population of valve lengths in a species of brachiopod can again be considered. An initial hypothesis could be that the lengths were approximately normally distributed. Probability theory indicates that if this were so, then the properties of that distribution could be described effectively from a random sample in terms of mean, variance or standard deviation, and if they are relevant, skewness and kurtosis, described in §3.10. The best estimate of the population mean that can be obtained from the sample is the sample mean, and likewise for the other parameters. The larger the sample, the more accurate is the estimate likely to be, and the degree of accuracy can be stated precisely if certain assumptions are made. In the ideal case of a truly random sample from a normal population, the standard deviation of the means of samples of n items is inversely proportional to the square root of n. Thus, if many samples, each of 100 items, were taken from a normal distribution, the standard deviation of the means of the samples, known as the standard error, would be $\sqrt{1/100}$ or $0 \cdot 1$ of the standard deviation of the population. Since the probability of a deviation of up to one standard deviation from the mean in a normal distribution is 68%, there is a 68% probability of the mean of a sample of 100 items lying within $0 \cdot 1$ standard deviations or one standard error of the population mean. There would be a $99 \cdot 7\%$ probability of it lying within three standard errors from the mean. The degree of certainty is often stated as confidence limits, giving, for example, the estimated value of the mean, and a range of values which is likely to be exceeded in not more than, say, five cases in a thousand. This degree of risk of error might be acceptable. If the conclusion were relatively unimportant, finer limits could be put on the value of the mean, with a greater risk of being wrong. On the other hand, if an important decision, such as the selection of a seismically inactive area for disposing of radioactive waste, was to be decided on the basis of the mean value, one might wish to make a statement that would be expected to be wrong only once in a hundred thousand times. Either a larger sample would then have to be collected, or a wider range of values quoted. The value of the mean might thus be stated in the output from a computer program as, say, 56 mm± $1 \cdot 2$ mm at the $0 \cdot 005$ probability level. Alternatively, knowing the mean, standard deviation and number of items, confidence limits for a chosen probability level can be looked up in statistical tables that can be found at the back of many statistical textbooks or in books of mathematical tables.

A number of points may be noted about the procedure. First, the results require modification if the sample is small, or if the sample forms a large proportion of the total population (see Stuart, 1962). Second, the so-called Central Limit Theorem indicates that the distribution of means of a set of large, random samples drawn from the same population is approximately normal, regardless of the form of distribution of the population. A sample of over thirty items could usually be regarded as 'large' in this context. Therefore, standard errors and confidence limits can be calculated for a large sample whether or not it is drawn from a normal population, provided the sample is a random one. Third, the standard error of the estimate of the population mean is inversely proportional to the square root of the number of items in the sample, provided that the sample is a comparatively small proportion of the population. It is thus the number of items in the sample, not the proportion of the population sampled, which determines the standard errors. Similar comments apply to the estimates of the variance and standard deviation, for which confidence limits can be calculated.

An important point that applies to many techniques of quantitative explanation is that deductions are based on knowledge of the population, not of the sample. Yet, in the nature of things, the available information concerns the sample, not the population. The mean, the standard deviation, and the frequency distribution may be known from the sample. If they were known in advance for the population, the investigation would be unnecessary. In the particular case of the parameters of a variate, some direct mathematical reasoning from sample to population is possible, but the quandary is more often resolved by careful formulation of statistical statements, and in particular, the use of the so-called 'null hypothesis'. For example, if it was hoped to demonstrate a difference in the lengths of brachipod valves at two localities, then a sample of measurements could be collected from each locality, the frequency distribution examined and the mean and variance calculated for each sample. A null hypothesis could then be set up to the effect that both samples were drawn from the same population, the parameters of which could be estimated by combining those of the two samples. If the null hypothesis were true, then conclusions could be drawn about the values of the means and variances lying within certain limits. If it was found that they did indeed do so, then there would be no reason to disbelieve the null hypothesis. If, on the other hand, they did not, this could be a chance occurrence, even if the likelihood is 1 in 1000 or less; or the samples might not be random; or the assumptions about the parent population might be in error; or the null hypothesis might be wrong. Under most circumstances, the last explanation could be taken as the most likely, and the conclusion therefore drawn that there is a significant difference in the lengths of the brachiopod valves at the two localities. Statistical inference can be a valuable guide to the credibility of results, and may indicate conclusions that are by no means self-evident. But it is also important to pay careful attention to the form of the reasoning, and to keep a watchful eye on alternative explanations of the results.

In a rather similar manner to means and variances, confidence limits can be calculated in correlation and regression analysis, and the null hypothesis can be invoked as before. Confidence limits have been calculated for trend surfaces, for instance, consisting of two surfaces, one below and one above the most likely surface (Krumbein and Graybill, 1965). This provides a good geological example of the need for careful formulation of statistical statements. Does the zone within the confidence limits indicate where the entire trend surface of the population is likely to lie? Is the population in question all possible points on the surface, or all possible surfaces? One can imagine that if present-day Arizona were buried under 10 000 feet of sediment, and a small number of boreholes sunk through the present-day land surface, that the Grand Canyon might be missed entirely in the sample, and in that area a rather flat trend surface might be established with close confidence limits. The canyons might lie many thousands of feet below that surface. A new unanticipated geological process had operated, and the assumption about a homogeneous normal population would be incorrect. Essentially there would be two types of surface in the area, with different geological origins and characteristics. The trend surface and its confidence limits could be useful descriptors for the desert surface, provided the geologist keeps in mind the thought, that, no matter how close the confidence limits may be, if the Grand Canyon does not appear in the sample, that does not prove that it does not exist.

The null hypothesis might also be invoked in examining a matrix of correlation coefficients. The hypothesis might take the form that each correlation coefficient referred to two variables randomly sampled from normal populations which were uncorrelated. Again, statistical tables could be consulted to determine, for the appropriate sample size and level of confidence, which values in the correlation coefficient matrix were not consistent with this explanation, and for which, therefore, some other explanation must be sought, such as non-representative sampling, strongly non-normal distributions, or a genuine correlation of the variables. As is always the case, however, a failure to disprove the null hypothesis does not imply a proof of it. Variables are not necessarily uncorrelated even if the correlation coefficient lies below the level of significance. It may merely be that the sample size is too small to provide clear evidence of the relationship. A careful examination of individual items in the original or the standardized data matrix might indicate whether or not this was likely. The null hypothesis is in a sense representing an expectation based on background knowledge. If there were no pre-existing reason to suspect that two variables were other than uncorrelated, a failure to disprove it would be of little interest.

4.5. Causality and correlation

The geological explanation of correlation between variables requires consideration of various possibilities before attributing the relationship to a geological cause. The concepts of cause and effect are intuitively familiar to geologists and are usually employed without ambiguity. The child who believes that the wind is caused by trees shaking their leaves probably has an inadequate understanding of the physical processes rather than problems with the philosophical doctrines of causality. But similar misunderstandings can arise in attempting to explain a matrix of correlations. The fact that one variable changes magnitude in unison with another does not necessarily imply cause and effect. The extent of movement of leaves on a tree and the wind velocity may be correlated, but a hypothesis about the underlying mechanisms is necessary in order to explain the correlation. Correlation between two or more variables may be due to a common cause, which may or may not be included in the data matrix. For instance, the grain size of sediment and the abundance of a certain type of fossil may be correlated because the same depositional conditions were favorable to the development of both. Many geological variables change gradually in space, and an investigation over a limited area may reveal a correlation between variables which merely reflects coincidence of trends over that limited area, as illustrated in Fig. 70. There are well known examples of similar 'nonsense correlations' due to trends in time, such as the correlation between the number of swimming pools installed each year in the United States and the incidence of lung cancer in Australia, or the annual frequencies of papers on mathematical geology and prescriptions for anti-depressant drugs.

4.6. Hypothesis testing

The framework of explanation arises from information outside the data, but may give rise to hypotheses that can be tested on the data. Hypotheses to be tested on a data matrix may lead to predictions about relative numbers in various classes. In these circumstances, the aim of the statistical test is to determine whether the discrepancies between the observed and the expected values are significant at a defined probability level. An appropriate test is the χ^2 test (pronounced kye, rhyming with eye, square). The observed value (O) is compared with the expected value (E) by the formula $(O-E)^2/E$ and

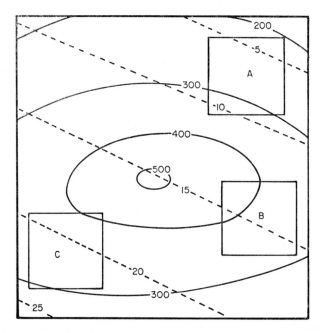

Fig. 70. Nonsense correlations between autocorrelated variables. The thickness of a wedge-shaped sand body is shown by dashed lines, the present-day topography of the same area by solid lines. Within the square marked A, there is strong correlation between the variables, within C a strong negative correlation, and in B no correlation is apparent. Both variables show strong trends, but there is no causal link between them.

this is summed for all classes. The number of items occurring in certain classes and the numbers that would have been expected had the categories been uncorrelated are tabulated in Fig. 55, and the value of χ^2 calculated on the differences between them. The value of χ^2 can be looked up in statistical tables, to determine whether there is reason to disbelieve the null hypothesis that there is no significant difference between the observed and expected values.

If a number of samples are drawn from a population that is more or less normally distributed, the distribution of sample variances can be predicted. The so-called F test is available to determine whether there is reason to disbelieve the hypothesis that two sample variances were drawn from the same population. By careful experimental design, as mentioned in the sections on sampling (§2.7) and data analysis (§3.16), it may be possible to devise an experiment in which a number of controls which affect the value of a variable interact in a known manner. Representative measurements of the variable under the influence of each combination of controls can then be compared. This is done by partitioning the variance of the variable among the identified controls, so that each is regarded as a possible contributing source of variation. The F test is used to determine which partitions of the variance are significant in terms of the null hypothesis, and hence to decide which controls of the variability might be important. Examples of this type of application were mentioned in §3.16. The F test can also be applied in a similar way in multiple regression analysis (§3.15) where a proportion of the total variance is associated with each of the coefficients in a regression equation, thus helping the geologist to decide which controls are significant in determining the value of the independent variable. Analysis of variance and regression analysis are thus analogous techniques, but with the controlling factors measured in terms of qualitative classes in analysis of variance, and as quantitative variables in regression analysis.

4.7. Computer simulation; random numbers

The success of statistical deduction in such fields as analysis of variance, is dependent on careful experimental design and appropriate sampling. But the geologist is often concerned with processes that took place in the remote past under conditions that are known only by inference, and were certainly not under the geologist's control. Even the sampling pattern is likely to be determined more by accessibility of material than by the geologist's decisions. Statistical inference, based on knowledge of frequency distributions and sampling patterns is very difficult under these conditions, and the geologist wishing to explain his data and the results of data analysis may have to resort to simulation methods, where there are fewer mathematical constraints and more room for geological intuition.

Simulation methods do not normally attempt to reproduce observed data values, since these include a random element which would not be duplicated except by chance. The objective is rather to determine the pattern of results that might be expected from a defined set of conditions, sometimes known as the scenario. The comparison with the observed data is through summary statistics derived by data analysis, such as the mean, variance, correlation coefficients, and even principal components or regression coefficients, since the statistics are likely to reflect aspects of the underlying processes with greater stability than individual values would. There may be some fixed aspects in the scenario, such as the position of the sampling points, which it is simplest to duplicate precisely. More generally, it is necessary to consider a range of variation both in the model and in its parameters, to determine which scenarios are compatible with the data. The conclusions are likely to have the same form as the null hypothesis mentioned above, namely, an indication of whether there is or is not reason to disbelieve a particular model on the basis of the available data. The acceptability of the model as an explanation depends on the framework of other knowledge within which the model is considered. It is to be expected that more than one scenario would be compatible with the data, and probably with background knowledge as well. Consideration of rival models might then point to additional data that could be collected to resolve the ambiguity. In any case, a wider range of models can be considered than the rather limited number that can be handled readily by mathematical analysis. An example may help to clarify the method.

Data are collected of the thickness of a sand body at boreholes at the locations shown in Fig. 71(a). The thicknesses, in feet, are shown on the diagram. Trend surface analysis, and visual inspection, suggest that this is a linear body with a northeast trend. One hypothesis that could be considered is that the values are in fact randomly distributed, and that the apparent trend merely reflects the pattern of data points. This hypothesis could be tested by generating random values at the borehole localities and determining whether the impression of a linear trend was given by any of several sets of random values. The two points of immediate interest are the technique of simulating the data values, and the method of testing the conclusions.

One straightforward way of obtaining random values would be by rolling two dice, thus obtaining a number between 2 and 12 with each number having an equal chance of being chosen. With each throw of the dice, a number could be entered at one location. The results of this procedure are illustrated in Fig. 71(b). It is preferable to enter the data in a systematic manner, such as starting at the north edge and working south, going from west to east where points are at the same latitude, thus following the conventions of the printed page. At first sight, entering the values in a haphazard manner would seem to improve their randomness, but in fact would increase the risk of unconsciously placing high values, say, at the centre, and low values nearer the margins. Apart from the geographical location of the data points, the results of this procedure have practically nothing in common with the observed values, and even as a null hypothesis, are somewhat unconvincing. The procedure of rolling dice gives rise to a rectangular distribution, so called from the shape of the frequency distribution curve which indicates that every value has an equal chance of being selected, within the given range. The value of zero has no chance of being thrown, and the mean and variance do not correspond to those of the sample of sand thicknesses. The simulation experiment is thus so unrealistic that it throws little light on the question of interest. Other objections to the procedure are that the mechanics of throwing dice become very tedious, and it is not possible to reproduce the same sequence of random values, as might be desirable if the experiment were to be repeated with perhaps the same random elements but a smaller number of data points. Much of the value of experimentation lies in the ability to hold

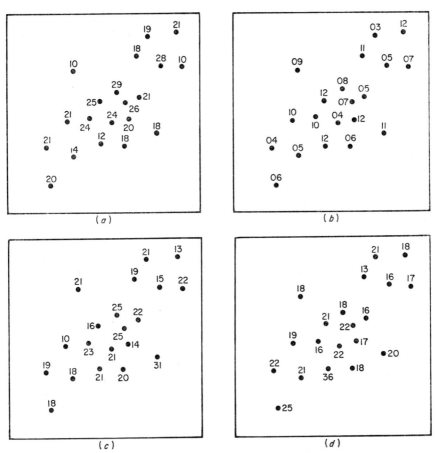

Fig. 71. Thicknesses of a sand body from boreholes. (*a*) The original data, possibly showing a northeast trend. (*b*) Random values between 02 and 12 from a rectangular distribution. (*c*) and (*d*) Two sets of random values from a normal distribution with, like (*a*), a mean of 20 and standard deviation of 5

one set of variables, including any random variation, at a steady value while introducing controlled variation in another set.

These problems can be overcome in a computer simulation by generating numbers with appropriate properties in a subroutine in the computer program. The design of the best algorithm for this purpose is itself a complicated subject. Computational efficiency is important, since long series of these numbers may be required, and the algorithm may be specific to a particular model of computer, in order to take advantage of hardware features. A possible procedure would be to take a specified 32-bit number, calculate its square to 64 bits accuracy, then drop the first and last groups of 16 bits, retaining the central 32 bits as the next number in the sequence. This number in turn can be the input to the subroutine to generate the next number. In this way, a sequence of numbers is generated, which is reproducible, since the same set of numbers would always be created from the same starting value. To this extent, the numbers are predictable and not therefore truly random. They are known more strictly as pseudo-random numbers or PRN's, but as the distinction is not important here, the term random number is used loosely to refer to PRN's.

Subroutines to generate random numbers are available on most computer installations. The basic subroutine generally provides a number from a rectangular distribution between $0 \cdot 0$ and $1 \cdot 0$, subsequent transformations being used to arrive at other distributions. An alternative routine, with the transformation incorporated, is usually available to generate random numbers from a normal distribution with zero mean and unit standard deviation. A subsequent transformation could provide numbers from a log–normal distribution. A set of values drawn from a normal population with specified parameters can be obtained by multiplying the standardized output by the standard deviation and adding the mean value. For instance, analysis of the data for the thickness measurements from Fig. 71(a) suggest that they might be a sample of a normally distributed population with mean of 20 feet and standard deviation of 5 feet. A set of random numbers from a distribution with these characteristics can be generated as shown in figure 71(c) and (d) from the output of a standard normal distribution. For a paper exercise of this kind, where the main calculation is not done on the computer, published tables of random numbers are available in books of mathematical tables and appendices to books on statistics. By these methods, a sample of values can be plotted on a map, which are known to have no geographical pattern or trend, but which are comparable in other respects to the observed data, and can be regarded as representing a null hypothesis. If the simulated data appeared to the geologist to show a pattern similar to the observed data, this would throw doubt on the reality of the trend, and would suggest that it could be an illusion resulting from the pattern of distribution of the data points. Because of the random component in the simulation, however, different simulation runs, based on different sets of random numbers, will give different results. More rigorous methods are therefore required for comparing the simulated and the observed outcome, and thus testing the hypothesis against the data.

4.8. Testing subjective judgment; retrospective prediction

One approach to such testing is visual comparison of the simulated and the observed data by an experienced geologist. The test is also of the perception of the geologist, and it is difficult to separate the two aspects. This is somewhat comparable to the testing of subjective opinion which is necessary, for example, in testing new drugs or in market research. In order to have confidence in the results, several geologists should be involved, and they should not know beforehand which is the observed and which the simulated data. A number of simulated outcomes would be necessary, in order to make allowance for the random element. A possible result of such a sequence of tests could be to show that a geologist might, for instance, consider that the real data exhibited a trend which he did not believe to be present in any of the simulated maps, although these were based on random numbers drawn from the same distribution and with the same parameters as the sample, and located at the same points. The simulated data can be elaborated by introducing trends of various types and strengths. For example, a linear trend, mimicking that which is thought to be present in the data, might be introduced by adding a manually generated set of numbers representing an idealized trend to the random values, taking care not to alter the statistical properties. A sequence of plots giving different weights to the ideal pattern and the random distribution might test the level at which the trend becomes apparent to the geologist.

The above procedures have distinct limitations, since they are testing the geologist's perception rather than leading directly to a geological conclusion. Nevertheless, since many decisions depend on this type of judgment, examples can be useful in helping the geologist to calibrate his impressions against known patterns. An alternative approach for testing a model is appropriate where the model is designed to have some predictive value. This is to apply the model to a selected part of the data, and use it to predict the values of the points which are known but not included in the test data set. The predictive power of the model can then be tested by comparing the generated values with the known data points which were not accessible to it. Returning to the example, a selection of half the data points might in itself have been adequate to indicate a trend, perhaps by trend surface analysis, which might provide an explanation of the distribution of the values in space, but could also predict probable values through the area as a whole. If the hypothesis has some predictive value it is not unreasonable to place greater confidence on the validity of the model as a result (see Harbaugh et al., 1977).

4.9. Interaction of variables; sensitivity analysis

The methods described above are most likely to be successful with relatively simple models. Where a wider range of information is available, it may be possible to construct a detailed quantitative model to assist in explaining the observed data. Simulation can again play a part in testing the model.

The first step is the calculation of summary statistics for the observed data, by the methods of data analysis and interpretation already considered. Similarly, the results of simulation runs can be summarized statistically. The objective is to compare the two sets of summary statistics to answer the question: Are the observed data within the range of possible outcomes of the simulation model? The ease of experimenting with a well constructed computable model means that individually defined scenarios can be considered. Where random variation is a significant part of the model, a rather large number of simulation runs may be needed to give a clear picture of the frequency distribution of the possible outcomes.

The essence of an experimental investigation is the ability to alter the variables in a controlled manner in order to determine their effects, singly and in combination, on the system with which the experiment is concerned. In simulation, the corresponding activity is known as sensitivity analysis. The objective is to determine which aspects of the computable model have the greatest effect on the simulated outcome, and the manner in which the outcome is affected. To achieve this, all aspects of the simulation are held constant, including the values of the random numbers, except for the variable on which the sensitivity analysis is being performed. The value of the subject variable is held at several selected levels in turn, and if there is a random element, a number of simulation runs are made at each level and the results tabulated. A variable which does not have a random component obviously requires only one run at each level. The response of the model to changes in a random variate can be studied, not only in terms of changes to the mean value, but also to the standard deviation, or to the form of the frequency distribution.

Sensitivity analysis is not confined to single variates, and the results of most interest may concern the interaction of two or more variates. Mathematical experimentation with the computable model then requires an experimental design similar to those used in more conventional forms of analysis of variance. The analysis has a similar purpose, namely to determine the contributions of the various controlling factors and their interactions to the variation in outcomes of the model. The geological aspects of the model must be carefully considered to ensure that the results are as relevant and fruitful as possible. The difficulties of experimental design in examining the results of a real geological process are not encountered in the computer model where the simulation process is under the geologist's direct control.

The methods of significance testing described earlier in this chapter may be relevant in determining whether the observed data are consistent with the scenario represented in the computer model. The tests still have the form of the null hypothesis, and do not normally provide positive proof. They might, however, indicate that at least there is no reason, on the basis of statistics derived from the sample of observed data, to disbelieve the hypothesis that the sample was drawn from a population with the properties of the data generated by the model. While this is not in itself sufficient grounds for accepting the hypothesis, it may be a necessary condition for giving it further consideration. If the objective of a geological investigation is to arrive at an adequate explanation of the geology of an area, then consideration of a range of possible models, whether or not these are quantified, may suggest where emphasis should be placed in data collection in order to test and discriminate between the models. The construction of models is thus a central activity in the process of explanation, and may have value as a means of presenting a set of ideas and testing their logical consistency, as well as a framework for exploring and testing the consequences of various scenarios.

THE SYSTEMS APPROACH TO MODELLING

4.10. Integration of ideas in a computer model

Explanation implies that observations and the results of data analysis are considered within a broader framework to clarify their meaning and significance. But the implications of raw data which may be familiar to the geologist are not usually explicit in the computer record. The fact that Cambrian rocks are older than Ordovician, or that granite is richer in silica than gabbro, for instance, is unlikely to be stored with the data. The geologist may also have background information relevant to explanation, such as the knowledge that in the Ercu Mountains the Cambrian strata are thinner and less extensive than the Ordovician. It is therefore necessary to consider how such implications and expectations can be introduced into the computer model. The raw data are also liable to be in a form which does not readily lend itself to direct comparison and explanation. For instance, a set of boreholes in an area might yield information about the elevation of a set of lithostratigraphic and biostratigraphic horizons at a number of scattered locations, and seismic surveys might give the depth of reflector horizons along traverse lines. The raw data from traverses and boreholes cannot be compared directly, and for the geologist, the interest lies in the surfaces and their relationships, rather than in the data points themselves. In making the comparison by computer, interpolated values can be generated using the same frame of reference for each surface, and comparisons made on the basis of the interpolated grid values. An interpolation algorithm is required which reflects the relationships inherent in the raw data. Such algorithms are an explicit statement of ideas that are normally implicit in geological thinking.

4.11. Analysis of the system

Consideration of how to make implicit information available to the computer program is a useful first step in building a computer model and is considered in more detail below. But the systems approach can lead to integration of ideas on a wider scale. It is basically an application of common sense methods that are appropriate for any complex problem, whether or not a computer is used. However, a formal approach can ensure that the model can be represented and studied on the computer in whole or in part. The systems approach to modelling is adopted in many subjects, as diverse as economics and ecology, where relationships are complex, and are described in tentative hypotheses rather than as firmly established laws. The first step in constructing the model is to analyze the system into its components and the relationships between them, such as the subsystems, the entities and their attributes, the activities and the events, processes, responses and constraints which characterize them. The division of the system into subsystems leads to modularity in the computer representation, which is necessary for the development of reasonably comprehensible and efficient programs. The explicit and precise statement of entities and relationships in the model is essential for computer representation. These aspects may absorb a large part of the total effort of model construction. In a model designed to have explanatory value, the emphasis is likely to be not so much on the raw data, but more on the hypotheses, theories and relationships which explain the data, and the programs and data structures in which they are embedded. The development of large-scale systems models can be seen as an important aim of mathematical geology, but they are still at an experimental stage, and the techniques described below have more immediate practical value in small, self-contained models. It is hoped, nevertheless, that their longer-term role in a more comprehensive model will be apparent.

4.12. Expressing implicit information;
table lookup and cross-reference; generated data

A simple example of transforming data in order to make implicit information available for computer processing is conversion of data to codes which have some semantic content. Although it is possible to use semantic codes (see §2.14) directly for recording raw data, it is usually more convenient to use a mnemonic code for data recording and input, and to convert this, by looking up an index on the computer, to a code which contains information, quite possibly at the bit level, in a form directly usable by the processing programs (see Fig. 25). The application can be quite straightforward, such as replacing a mnemonic chronostratigraphic code with a numeric code which indicates relative position in the stratigraphic table, or even indicates an approximate age in megayears. This immediately makes it possible to sort the data by computer into correct chronostratigraphic order, or to plot data values against time. A slightly more complex example would be to replace rock-names in a data file with a semantic code which indicates the petrographic class, or even the possible geochemical content of the rock. If the data were then searched for all rocks with over 10% of silica, records of granite could be picked up through the semantic code. The same result could have been obtained by searching the rocknames directly for terms such as granite, as well as searching the geochemical analyses. With complicated tasks, however, such as assembling and printing all data records arranged according to their geochemical content, there can be significant advantages in having the necessary information incorporated in each data record. This does not necessarily mean that the full code should be stored permanently with each record, however. Provided the lookup table is available, the code can be added at processing time. This introduces additional flexibility, since the code can be related to the application. For instance a link between the rock name and the geochemical characteristics might be appropriate for one application, a link to a table giving the properties of the rock as a building stone might be appropriate for another.

The semantic code can give a very compact representation of information. The reducing cost of computer storage, however, has meant that the advantages of compactness are generally outweighed by the complexity of processing the coded information. An alternative means of expanding the data content of a file is to link it to a second file which contains in explicit form the information implied in, say, a rock name. By combining data from the two files, additional data fields can be added to the original data records for processing purposes.

Some information which is implicit in the data can be made explicitly available by calculation rather than by table lookup. A simple example is the generation of thicknesses of stratigraphic units from a file which contains only depths as raw data. With more complex examples, the method of generating data is less clearly defined, and may involve a choice of hypothesis or model. An example is provided by silicate recalculation or normative analysis, in which the geochemical analysis of an igneous rock is the basis for calculating a normative mineral composition, which could be a means of comparing the rock with those known only from petrographical examination. Several procedures are available for silicate recalculation, each of which may give a slightly different result. Most methods are based on those originally designed for hand calculation. In one particular method, the chemical oxides present in the original analysis are removed in the ratio appropriate for the first mineral to crystallize from the magma. The removal stops when one of

N

the oxides forming the mineral is exhausted, thus determining the proportion of that mineral. The extraction of the second mineral in the crystallization sequence is then simulated until one of its component oxides is removed from the mixture, and so on until all the oxides are used up. One of the advantages of this method is that, although it is somewhat laborious to handle manually, it is not computationally difficult. With methods devised specifically for the computer, however, it is practicable to perform more complex calculations. The problem of silicate recalculation can be regarded as one of a class of allocation problems, and is considered below in more detail as an example of a geological application of an important modelling technique, see Harbaugh and Bonham-Carter (1970).

4.13. The allocation model; linear programming

One method for handling the allocation problem on the computer is known as linear programming. A limited supply of components are combined in different proportions to give products of different compositions. The objective is to determine the quantity of each product. The availability of the components is an important constraint, but does not itself define a unique outcome. The other item of information required is the so-called objective function, which describes the behavior of the system in allocating components. It measures the degree of preference within the system for the use of each component and the preparation of each product. In a commercial example, the system might be a factory which bought components for a particular price, and manufactured a range of products from them, each with a market value. The objective function in that case would reflect the cost of the raw materials, and the price obtained for the manufactured goods, and would be maximized to give the highest profit. A geological example is a molten magma in which the constituents are a number of chemical oxides in defined proportions. As the magma cools and crystallizes, the chemicals combine in various proportions to form minerals. The amount of each mineral in the solid rock reflects the availability of the constituents. This is one clear constraint on the system, since the amount of quartz clearly cannot exceed the amount of silicon and oxygen in the magma. However, this constraint is not sufficient to determine the behaviour of the system, since many combinations of minerals could be produced from the same proportions of elements.

The fact that some minerals develop in preference to others is well known to the petrographer, and is implied by the order of calculation of silicates in normative analysis. In linear programming, this preference is expressed by the objective function which associates a value, analogous to profit, with each possible product of the system. The value of the objective function for the system as a whole is calculated by multiplying the quantity of each product by the 'profit' associated with it, that is, the value of the product minus the cost of the components, and adding the results for all products. Linear programming optimizes the objective function for the system. That is, it calculates the amount of each product which will completely use the original components, and will give as high a value as possible for the objective function. If, for example, the model assumed that as much olivine as possible would be produced, then this would have a high value in the objective function. If quartz were produced only if free silica remains after as much as possible of the other materials has been produced, then quartz would have a low value in the objective function, and so on.

The parameters of the objective function could be obtained by trial and error. Working from instances where the chemical and petrographic composition are both known, the parameters could be chosen to ensure that the method gave reasonable results. This could then be extended to other rocks for which the petrography was not known, provided that the method was constantly checked by examining the actual petrography of a sample of rocks, and if need be modifying the parameters to improve the results. This 'heuristic' or 'learning-by-doing' approach will be mentioned again in other contexts, but is not in itself a completely satisfying approach to determining the objective function, since it lacks explanatory value. The complementary approach is to consider the geological hypothesis which the model represents, and see whether there is theoretical or experimental guidance to the suitability of the model and the form of its parameters.

The first point to note is that the two methods of silicate recalculation described above represent different geological processes. In both cases, minerals crystallize from a chemical melt with preference being given to the formation of some minerals rather than others. By extracting normative minerals in a specified sequence until one of the necessary oxides is exhausted, one can model a crystallization process in which order of crystallization in a cooling magma determines the preference given to the formation of each mineral, which withdraws its constituent oxides from the magma until the supply of one is exhausted. In the simple case, that mineral is then no longer an active part of the system, which is thereafter considered as though it were free of that component. The second approach, of allocation by objective function, assumes that there is a chemical melt in which all the normative minerals have an opportunity to compete for the available oxides, but with some better able than others to attract the chemicals they require. Looked at in this way, some physical significance can be recognised in the preference for certain minerals, perhaps reflecting the temperature of crystallization in the first model, free energy in the second. It is not the purpose here to recommend a procedure for normative analysis, but simply to point out that different models can be constructed to reflect different hypotheses, and that by linking them to the hypothesis, their relative merits can be considered in particular instances, and meaning assigned to the parameters of the model. The complexity and variety of geological processes suggests that a single model is unlikely to meet all cases, and that several rival models may well be required to represent the same process in different circumstances. This aspect can perhaps be illustrated by examples in the field of spatial analysis.

4.14. Contouring; the spatial model

There are several methods of drawing contour maps by computer for a surface known only at scattered data points, such as boreholes (see Harbaugh et al., 1977) A number of interpolation algorithms are available, each of which reflects different ideas about the behavior of the surface. Background knowledge of the geology or the sampling pattern may be helpful in selecting an algorithm and choosing its parameters. A rather extreme example is where a surface, such as present-day topography, is known in detail, and contours are already available for digitizing from a map. In order to avoid storing an excessive number of data points in the computer, the digitized values might be chosen to ensure that a close approximation to the original contours could be recreated by the computer. Where the contour line is straight, it can be recreated from two points; where it is a smooth circular arc, it can be recreated from three. Provided the mathematical procedure for generating the curves is known, the draftsmen who are digitizing the line data can be trained to select points which will enable the original lines to be accurately reproduced by the interpolation algorithm. The algorithm as well as the data are then essential parts of the computer model of that surface, since, with a different algorithm, the original contours would not be reproduced.

Many contouring methods require, as an intermediate stage, the calculation of values of the surface at regularly spaced grid points. The grid values, see Fig. 72, are calculated by interpolation from the surrounding data points. A simple method of interpolation is to take the average of all data values within, say, one mile of the grid point. This gives poor results if the data are not spread evenly across the area. The data points might, for instance, be spaced closely over oil fields, and widely scattered elsewhere. They might, particularly with geophysical, oceanographic or geochemical data, be densely spaced along traverses with little or no data between the lines. In these circumstances, interpolation to a grid point might be strongly influenced by a cluster of many points a mile away, while the few data points near the grid point would have comparatively little effect. With some data, such as geo-chemical analyses of stream sediments, there may be much more local than regional variation, and the average value would be suitable for interpolating

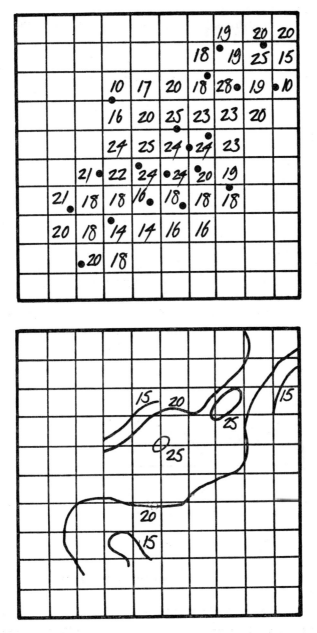

Fig. 72. A surface, such as that of Fig. 71(a), can be represented digitally by grid values or by digitized contour lines. The grid values are easier to handle mathematically and can be used to generate contours.

to the grid point. With some other types of surface, like the elevation of a smooth peneplained disconformity, there may be little likelihood of rapid change in surface level over a short distance, and the data values close to the grid point should be weighted more strongly than more distant points. Another form of interpolation is then required, in which the contribution of each data point to the average value is inversely weighted by the distance, or more commonly the square of the distance of the data point from the grid point.

Traverse data can impose an unwanted directional element to the interpolation. A set of closely spaced points does not necessarily give more information than fewer points spread more widely. If the surface is fairly smooth then closely spaced points are in effect giving the same information many times. In order to compensate for this, interpolation algorithms may consider separately eight octants around the grid point and weight the values in each octant so that all eight play an equal part in determining the grid value (see Fig. 73).

The probable form of the surface can affect the choice of algorithm in other ways. With a regional field, say the gravitational or magnetic field over an area, it may be the broad overall changes rather than the local fluctuations, that are of interest. There could then be theoretical grounds for basing the grid values, and hence the contours, on a high-order polynomial trend surface, which would take into account all the known data values throughout

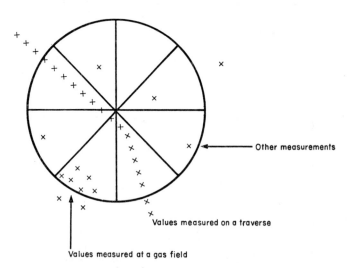

Fig. 73. The depth is to be interpolated at the center of the circle. Only points within the circle are considered, and each is weighted by the inverse of the square of the distance from the center. To counteract the uneven distribution of data points, those in each octant are considered separately, and the octants are weighted equally in calculating the final value.

the map, rather than considering each local area separately. At the other extreme are geological surfaces with very abrupt local variation, such as a heavily faulted area, or a surface with biohermal reefs, or salt domes, or canyon topography. For these surfaces it would be undesirable and misleading to take an average value from a wide area around each grid point, as this would have the effect of smoothing what are known to be sharp discontinuities. A more realistic contour map might be obtained by restricting the area for interpolation, or even by searching for discontinuities and zones of rapid change, and considering them as barriers to interpolation. A careful choice of model which takes background knowledge into account can do more to improve the final contour map than additional data would. The choice of parameters can be important as well as the choice of algorithm. In a large-scale study, there is a case for assembling quantitative information, such as the scale of variation of surfaces of various types, on a systematic regional basis. This could serve as a guide to the most appropriate parameters and techniques to apply to a particular requirement. Not only can data be exchanged, but also the models which give them meaning.

The spatial model, whether used for automated contouring or for other methods of spatial analysis, is worth considering further as an illustration of the possibilities of integrating ideas from various sources in a single computer model. The form of the model depends on the objectives of the investigation. A contour map which is drawn manually is seldom prepared with a single explicit objective in mind. One objective might be to estimate the probable elevation of the surface at a location where no direct data was available. This might be in response to a question such as: 'At what depth would a borehole at location (x, y) be likely to encounter Carboniferous sandstone?' A second objective might be to estimate the volume below a surface. 'What is the volume of coal shown by isopachs of the coal seam in area (a, b, c, d)?' A third objective might be to investigate whether a surface with known properties could be realistically fitted to the data values. 'Can a surface, represented by contours, which would be consistent with the form of an erosion surface developed by a system of river valleys draining to the southwest, be drawn to fit the data points in the area (a, b, c, d)?' Perhaps the most usual objective, however, is to give the geologist an overall impression of the surface, to enable him to visualize the overall form and shape, and to see the nature, scale and pattern of features and irregularities superimposed on the overall form. In automated contouring, a precise definition of the objectives may help in selecting the most appropriate method, even if the final choice involves some compromise.

For some of the objectives mentioned above, it is possible that the most appropriate computer method would not involve contouring at all, but would rely on other methods of spatial analysis. The most accurate procedure for estimating the elevation of a surface at a specific point may be to interpolate directly from surrounding data points. The simplest method of calculating the volume between a surface and a datum plane may be from the average height of the surface over the area of interest. If, however, the properties of the surface, rather than of individual points, are of interest then other aspects must be taken into consideration.

A variable which has a defined value at every point in an area, can be considered in geometrical terms as a surface. Many geological surfaces, such as the top of the Cretaceous limestones in Ercu County, have a real physical existence in space. Other variables which do not, can still be represented as surfaces, such as the thickness of the limestone, a sand–shale ratio or average porosity. The same methods of spatial analysis and display, including contour mapping, are appropriate. By considering the surface, rather than the set of points, emphasis is placed on the overall pattern of the distribution in space, and the description and manipulation of that pattern. Values of individual data points are then of interest, not on their own account, but in the context of their surroundings. The spatial model has a central position in much geological explanation, as individual observations can, in many cases, be explained only within the three-dimensional framework of surrounding rocks, or the pattern of processes by which they were formed. This would apply, for example, to the third objective mentioned above, namely, to determine whether a surface with particular properties would readily fit a set of known data points.

Some techniques concerned with the description of surfaces were mentioned previously, including the following:

1. Calculating a regional trend surface, that is, a surface which is mathematically simpler and smoother than the real surface, and gives a generalized approximation to its form and position in space. The small-scale local variation, or residual, is considered as a separate component.

2. Spectral and autocorrelation analysis, which describes the scale of variation of the surface, in terms of the relative importance of components of different wavelengths.

3. Slope analysis, which considers the relative frequency and direction of slopes on the surface.

4. Discontinuity analysis, in which abrupt breaks, such as faults, that interrupt the more regular changes in level and slope of the surface are analyzed.

Such techniques are concerned with analysis of a surface, and determining its mathematical and geological properties. This could be of value in indicating how the surface was originally formed, if comparable mathematical properties can be determined for surfaces known to have been formed by defined geological processes. For example, the geometry of sand bodies can be related to their origin as sand bars, filling of river channels, etc; the geometry of a folded surface may be related to its deformation by structural movements; the shape of an unconformity or a landscape may depend on whether it was eroded by rivers, glaciers or the sea.

If a surface of unconformity, known at points where it was penetrated by boreholes, were analyzed by a geologist using conventional, manual methods, he might begin the analysis by sketching in rough contours to approximately delineate the form of the surface. The next step of positioning the contour lines more precisely might consist of mechanically spacing the lines so that they were roughly parallel, and evenly spaced with regular variation in the spacing. At this stage, the geologist may recognize pattern in the form of the contours, perhaps the dendritic pattern of a river valley and its tributaries, and might modify the position of the contour lines to follow the features of such a pattern, while still conforming to the known data points. If the result is convincing, with a reasonably simple set of contours fitting both the concept and the data, then it might be put forward as an interpretation of the surface, and taken as support for the implied explanation.

The techniques by which a geologist contours manually, moving on an iterative step-by-step basis from broad initial assumptions to more precise hypotheses, are not necessarily the most suitable for computer methods. It is not easy, for instance, to automate the process of fitting a dendritic stream pattern to existing data points. The human mind is skilled at postulating the nature of a process from observation of the response, and the geologist may be able to guess at the possible overall form of a surface, and the processes that formed it, by examination of a set of data points. There is little point in using the computer to do what the human mind can do better, nor in restricting computer methods to mimicking the conventional approach.

The usual technique for automated contouring is based on the calculation of interpolated values on a rectangular grid, as already described, with the geologist controlling such aspects as the size and shape of the grid, and the weighting functions for distance and direction. The surface is assumed to be smooth and regular on a small scale, and the contour lines are proportionately spaced between the grid points, generated by an algorithm which ensures that they are smooth with few abrupt discontinuities. The position where the contoured levels cross the grid lines is calculated by interpolation. The algorithm which is commonly used to fit the lines between these points is a mathematical function known as a 'spline', named after a device used by draftsmen to draw a smooth curve through a set of points. A typical splining function is a cubic polynomial, which is fitted to a small number of points at a time. Having obtained a best fit to, say, the first seven points, the first point is dropped, and the next in the series is included for calculation of a new polynomial. In this way, the line is extended through the required points, step by step. The contour lines generated by the spline function do not necessarily pass on the correct side of every data point, although they should fit the grid points perfectly. However, this procedure does ensure that the contour lines are smooth, regularly spaced, and are a reasonable fit to the data.

The procedure described above corresponds approximately to the first phase of manually contouring a surface, and is appropriate either for a rough working map, which the geologist can refine manually to express his own ideas, or as a generalized representation of the surface, where the geologist has no single specific hypothesis in mind. A surface which is generalized to this degree has properties which differ from those of real geological surfaces. If the data points are a random sample, it is unlikely that the highest and lowest points on the contour map are as extreme as those on the real surface. The smoothing which took place during contouring biases any estimate of the range of values. Well data from hydrocarbon exploration will give a particularly unreliable estimate of low values on a stratigraphic horizon in an area where anticlines are the best oil prospects. The contour map can therefore be seen as an interaction between a set of data and a hypothesis, with the advantage in the computer method that the hypothesis is explicitly stated. Methods based on trend-surface analysis, moving averages, two-dimensional splining and autocorrelation analysis might be considered as alternative models for generating the surface. Some program packages (see Harbaugh et al., 1977), include these as options for generating grid points which are then contoured as described above. A well presented map with clearly labelled contour lines, highs and lows, and geographical coordinates, makes the final product much more acceptable to the geologist, and although of little mathematical interest, does involve complex programming. Several commercially available contouring packages meet these requirements adequately, and some aid the visualization of the surface by also plotting perspective views.

4.15. Sequential data and the transition model

The move from considering individual data points to considering the surface on which the points lie is an important step in integrating ideas, and bringing together data and hypotheses. The next step in integration is to consider the interactions and relationships of a sequence of surfaces, as a geologist does manually in studying the three-dimensional framework of an area. It brings together a wide range of geological information in the computer. To handle it satisfactorily requires not only the ability to describe and compare surfaces, but also the ability to analyze vertical sequences of strata as known from boreholes and tie the results into knowledge of the lateral variation. This is a central task in subsurface geology, and involves large amounts of data as well as complex sets of hypotheses about the relationships among the data. The next component in the spatial model is thus the analysis and explanation of vertical sequences.

The information available from a vertical sequence of rocks, observed at outcrop, or penetrated by a borehole, may be qualitative, such as stratigraphic tops picked by the geologist; semiquantitative, such as the abundance of fossil species recorded by the paleontologist; or quantitative, such as geophysical well logs. Various methods of summarizing and analyzing the data were mentioned earlier, of which transition probabilities for qualitative data, and serial correlations for quantitative data were seen as being particularly important. They are important not only because they offer a clear mathematical notation for describing the sequences, but also because the notation is fruitful, in the sense of forming a basis for models explaining the relationships. Transition probabilities provide a means of comparing the properties of measured sections with those of Markov chains, which are mathematical sequences in which each value is partly random and partly predictable from one or more of the preceding values (see Schwarzacher, 1975). It is thus possible to determine whether there is evidence that items in a lithological sequence are related to those one or more units below. Serial correlation also throws light on whether sequential data follow a predictable pattern.

From a geological point of view, it is unsatisfactory to study the properties of one vertical sequence in isolation, since the properties of the stratigraphic units in the surrounding areas are likely to throw more light on its geological origin than are the strata immediately below it. In graphic representations, whether prepared by hand or by computer, the variation in a sequence over an area can be examined with fence diagrams and serial cross-sections. The first objective is generally to build up a stratigraphic framework by correlation of horizons across a number of sections. The stratigraphic correlation is based on recognition of similarities of pattern among a number of vertical sections. The similarities are obscured by random variation and also by systematic changes in characteristics across the area. The systematic changes are likely to have a geological explanation.

Stratigraphic correlation from geophysical well logs requires the handling of large amounts of complex quantitative data, probably available in machine-readable form, and it is in this area that correlation by computer methods has been explored in most detail. The same approach can, however, be adopted with other kinds of raw data, such as lithological logs. There are three aspects to the stratigraphic correlation. One is the reduction of the large amount of data that is contained in the logs for each well, and its reconciliation with information from samples and cores. A second is comparing the reduced data among a set of wells in a manner which makes geological sense and takes background geological knowledge into account. The third is the reconciliation of the correlations between vertical sections with the implications for individual surfaces and their behavior.

The first aspect is returned to below in considering the heuristic approach. The second aspect requires the development of suitable functions to describe the geological changes that might be expected between different sections through a vertical sequence. A starting point is therefore a consideration of geological explanations for lateral change in a sequence, and their representation as a computable model. Fortunately, some quite realistic geological changes can be represented by simple mathematical functions. Change in thickness between two points in a section can be represented by simple multiplication. The changes are likely to be related to lithology, with sandstone beds, for instance, possibly increasing in thickness at a faster rate than shale. To take this into account, the function has to operate on two variables, the thickness of successive units and their lithology. In comparing two sections, some beds may be totally removed, possibly with erosion of the underlying strata. In a sense, these minor disconformities and discontinuities can be regarded as negative deposition, and as with deposition, the amount of erosion is related to the lithology of the beds removed. The relationship is somewhat different to that of deposition, however, since although ease of removal is correlated with decreasing grain size in the coarser grades, the cohesiveness of mud may make it more difficult to remove than fine-grained sandstone. The nature of the eroded material is, of course, not known from observational data, since it is no longer there to be observed. The technique thus requires the geologist to contribute considerable detail about the hypotheses represented in the model. Another type of change is the lateral transition within a stratigraphic unit from one material to another, such as sandstone to shale, with a corresponding change in thickness. A function which describes this alteration must also refer to the two variables of lithology and thickness.

The functions described above provide a notation which enables the geologist to describe, not his observations, but rather his detailed ideas of how the relationships which he observes could have arisen. He is, essentially, describing a lateral transition function which relates the sections encountered across an area. The function can be expected to alter, not only across the area, but also from one stratigraphic level to another. Moreover, since it represents underlying geological controls which must have had continuity in time and space, the change in parameters of the function and the nature of the function should show a meaningful pattern. If the functions are considered throughout an area, they themselves define surfaces. It may thus be possible to bring together the analysis of spatial pattern required for the preparation of contour maps of a geological surface, and the analysis of lateral stratigraphic variation required for the preparation of cross-sections. The spatial model can thus be extended to the sequence of layers that are the known geological surfaces in an area.

4.16. Linking sequences and surfaces

In considering a surface, the geologist is concerned with the pattern in space of a variable resulting from a set of geological processes, and his attention is concentrated on the geometrical aspects of the surface. When he considers a set of cross-sections, he is concerned with lateral variation and continuity of the pattern in a vertical sequence. The entire three-dimensional framework can be visualized as a stack of surfaces at different levels, and the emphasis is again placed on geometrical pattern, and on the relationships between surfaces at various levels. A first step in studying these complex relationships is to bring the surfaces, and thus the information from the data sets, to a common frame of reference. The sets of raw data, although they refer to the same area, are likely to have come from different sets of locations. Higher stratigraphic surfaces, for instance, are likely to have been penetrated by more boreholes than those at lower levels. The depth of seismic reflector horizons will be known at points along traverses, not necessarily at borehole locations. One readily available frame of reference, which can be made the same for all the variables is the interpolated data at the set of grid points. Statistically, this is somewhat unsatisfactory, since the manner in which the raw data were manipulated to generate the grid points affects the results in a way that is not easy to predict. The difficulty, however, is inherent in the data rather than in the method, and must be borne in mind in coming to any geological conclusions. When the third dimension is considered, the same dilemma presents itself, and the available frame of reference for comparing vertical sections, which also ties in with the sets of surfaces is again the values at the interpolated grid points.

Some contouring and spatial analysis packages have the ability to present the gridded data as maps, cross-sections or perspective views. In practice, where a number of carefully examined and widely scattered deep boreholes are available in an area, the primary framework of stratigraphic correlation may be established from these wells. Other wells, and later the interpolated grid points can successively be tied into the framework to give a more complete, though more conjectural, overall picture. The procedure for building up a computer record of the three-dimensional framework, might well start with the preparation of isopach maps for each layer in turn. If lithologies are known, they can be displayed graphically by computer, and this may suggest suitable lateral transition functions. The program package is likely to include facilities for surface to surface operations on the gridded or on the raw data. Thus data for the thickness of reservoir rock and data for the average porosity can be multiplied together to generate the data for a porosity–foot map; or data for thicknesses of one layer can be added to elevations of the layer below, to generate data for a contour map of the higher surface. The layer-by-layer addition may give more consistent results when seen in cross-section than would values derived by separate contouring of each layer, which are liable to produce intersecting surfaces because of the mathematics of the interpolation. The three-dimensional grid built up in this way depends on the geologist's interpretation and cannot be readily and directly recreated from the raw data without following precisely the same algorithms. It can be satisfactorily developed only by a geologist or group of geologists and geophysicists who are thoroughly familiar with the area. The final product is a detailed model reflecting their explanation of the structure and correlation of that area. Once the detailed grid is stored, maps and cross-sections can be produced as required to display the interpretation, and the grid can be updated by following the original algorithms when new data are obtained from additional drilling or survey.

4.17. The heuristic approach; the classification model

The lithological properties of the various layers may be known directly from cores and samples. In an area of hydrocarbon exploration, the main information is likely to be from sets of geophysical well logs. They provide quantitative measures of properties of the rocks which indirectly reflect such features as lithology, permeability, porosity and fluid content. As each of twenty logs may be digitized at six-inch intervals down a 20 000-foot well, very large amounts of computer-readable data may be available. A primary requirement is thus to reduce the data to manageable amounts, appropriate to the task in hand. A rather similar problem is posed by the remote sensing data obtained from digitally recorded scans from aircraft and satellite-borne sensors.

The data from remote sensors may consist of six or more channels representing filtered bands of the wavelengths of electromagnetic radiation to which the apparatus is sensitive. The intensity levels for each channel are recorded for small areas of the earth, known as picture elements or pixels, as the device scans repeatedly across the path traversed by the remote sensing device. The objective of analyzing the data may be to determine which combinations of values have a particular significance, such as rivers, desert, cultivated fields or forest. Mathematically, the problem is comparable to that of relating patterns of values on geophysical logs to the aspects of the lithology which they reflect. One approach used with remote sensing data is the so-called 'heuristic' method, in which the scientist provides limited initial information, from which the computer extrapolates as far as possible. The geologist reviews the extrapolated information and corrects or refines it where he can do so from his own background knowledge. The computer program then amends the extrapolation algorithm to take account of the new information and performs a revised extrapolation. The process continues until the computer provides a classification of the entire data set which the geologist regards as satisfactory. The computer program has then combined the information in the data set with the geologist's special knowledge in a manner which makes highly productive use of the geologist's time, and has reconciled differences between the computer classification and the geologist's knowledge by automatically amending the algorithm or its parameters.

The repeated interaction which is necessary in this activity requires a graphic display system with which the geologist can interact directly, and rapid response from the computer. For instance, on a visual presentation of a remotely sensed image on a cathode-ray tube, the scientist might indicate with a light pen typical areas of sand, forest and prairie. The computer might set up an initial classification algorithm from this data, and respond to the scientist by indicating by color or symbols which pixels were classified by its algorithm into these three categories. On examining the results, the scientist might find, from his own knowledge of the ground, that an area of marshland was wrongly classified as forest. He could indicate typical areas of this new category with the light pen. The revised display might on examination be found to show areas of known prairie as marshland, and additional areas of 'ground-truth' would have to be indicated with the light pen. On each iteration, the computer is using the larger sample of known points to revise the parameters of the classification algorithm, and reconsidering the raw data on this basis. A similar procedure could be followed with geophysical well log data, where the equivalent of ground-truth is provided by cored intervals or intervals that are particularly well known from examination of cuttings. Where a large amount of data is available, typically in geophysical, remote sensing and text data, an alternation of this kind between specific information supplied by the user, and generalization from this to the larger data set by the computer, with the human being refining and the computer expanding the results, can be an effective procedure. It virtually requires interactive computing, which with a large data set, can be expensive.

4.18. The process–response model; cybernetics; simulation

The examples just described are concerned with the assignment of data to classes, and the extension and development of a classification as the result of information organized by the computer. The geologist, however, may have to explain the origin and genesis of his data as well as their present distribution; the processes and mechanisms which caused the observed state to develop, as well as the relationships and the static framework into which his observations may fall. The process–response model which may be appropriate for investigating these dynamic relationships has been more fully developed in areas like ecology and meteorology, where both the process and the response can be observed, and even subject to experimentation, than in geology where the processes of interest may have taken place in the remote past or deep in the earth's crust. Nevertheless, the geologist does deal with processes than can be observed, and even without direct evidence, can study events of the geological past. The process–response model is therefore relevant to his work.

Perhaps the simplest form of process–response model is the regression equation. A single response, the variable on the left-hand side of the equation, is related to one or more process variables on the right-hand side. The regression analysis procedure provides a means of determining the best parameters to represent the relationship. An extension of the method, known as canonical regression (see, for instance, Kendall, 1961) makes it possible to relate several response variables with several process variables. Essentially, any such model is a set of equations giving a mathematical representation of a process which transforms an input to produce a response. A set of regression equations on their own is generally inadequate to enable the geologist to organize a complex set of ideas about a geological process. Other concepts that were developed in cybernetics, systems studies and simulation may help.

A convenient starting point may be to think of a mechanical analog of the process. If the essential features of the process can be thought of in terms of a machine, then it may help in representing them on another machine—the computer. If the process of transportation and deposition of sediment can be thought of in terms of pumps, filters and valves, or the process of folding rock in terms of springs and dashpots, then the physical properties that should be included in the model may be more obvious. This approach might also suggest analogies with models that are already known and tested in other disciplines.

Cybernetics is concerned with control mechanisms in machines, organisms and other complex systems. A simple example is the governor which regulates the running speed of, say, a diesel engine which is generating electricity. If the load on the engine increases, the resulting decrease in speed is sensed by the governor, which adjusts the throttle setting to bring it back to its intended speed. Similarly, a warm-blooded animal can sense a change in its temperature, and modify its behavior, perhaps by sweating or shivering, to restore its normal blood heat. The information derived by monitoring the process, which is the basis of the self-adjusting mechanism, is known as feedback. Where a departure from normal conditions results in action to restore equilibrium, negative feedback is said to be operating, and the system is stable. There is positive feedback when a departure from equilibrium does not bring about efforts to restore the status quo, but instead increases the tendency to depart from the original state. The system is then unstable, and any perturbation or chance irregularity will cause a change of state. An example is a supercooled liquid which crystallizes as soon as suitable nuclei are present. It is possible for stability and instability to coexist on different scales within one system. A set of dunes on a desert or beach may be stable when viewed on a short time scale, in that any modification to the form of the dune by a passing animal, or wind from an unusual direction, would be quickly repaired when normal winds resumed. On the other hand, the dune system as a whole may be in constant motion, and possibly out of balance with the sand supply, so that over a longer period the system is unstable.

Many other examples of feedback can be seen in geological processes. Catastrophic events, such as mud flows, land slips, slumping, turbidity currents, earthquakes or volcanic eruptions display positive feedback. The system remains stable during a period while the conditions build up to the point where failure occurs. The onset of movement in the above examples triggers further movement. Dynamic friction is less than static friction, and channels created by the incipient turbidity current, or by lava forcing a passage through cooler rocks, allow the process to continue at a faster rate until, with the dissipation of the potential energy, there is a return to stability. During deposition of a bed of fine-grained sand, the pattern of deposition may persist for a prolonged period, because finer or coarser sediment does not fit the small-scale geometry of the bed surface and is therefore more exposed to the agents of transportation. Negative feedback would thus tend to encourage continuation of deposition of a narrow range of grain sizes on that bed. The situation is somewhat analogous to an area of climax vegetation where existing plants tend to prevent the establishment of competing species. A change in circumstances, such as a brush fire or a change in climate in the case of vegetation, or a change in the current regime or sediment supply in the case of the sand bed, triggers the change to another state, which may in its turn achieve stability.

A recognition of control mechanisms is important in building a simulation model of a system. A number of programming languages have been created with simulation in mind. DYNAMO provides an example of a language in which feedback effects can be readily represented (see Forrester, 1961).

o

Simulation languages fall into two broad classes depending on whether they are designed to represent continuous or discrete processes (see Harbaugh and Bonham–Carter, 1970). GPSS is an example of a language for simulating continuous processes, appropriate to some applications in hydrogeology and reservoir engineering where relationships can be expressed as differential equations. SIMSCRIPT is an example of a language for simulating discrete processes, where the system is regarded as having a number of separate, identifiable states. This approach is probably more widely applicable in geology. Although the use of a separate language is not necessarily helpful in geological applications, the terms and methods which they employ provide a means of expressing the basic ideas clearly and consistently. Subroutines are available in the main scientific programming languages, such as FORTRAN and PL–1, for many of the simulation functions, including generation of random numbers, and there are languages like SIMULA which are designed to handle simulation within the framework of a general purpose language.

Systems are described in different terms in different simulation languages. Typically, however, a system is regarded as consisting of a set of 'entities' with certain 'attributes'. The entities interact with 'activities' under certain 'conditions', creating 'events' that change the 'state' of the system. Thus the beds of sediment along a shoreline might constitute the entities of a system with attributes such as location, extent, grain size, and sorting characteristics. These entities might interact with activities such as sediment transport, under certain conditions of sediment supply and wave energy, creating events of erosion and deposition which would change the state of the system. The definition of components required in building the model is part of the systems approach. The systems view is broader than that of simulation modelling, however, the latter being restricted to certain subsystems. In addition to considering how the system of interest is related to the larger systems of which it forms a part, and to separate, parallel systems, the systems approach also leads to a consideration of how the system can best be subdivided into subsystems, how they fit together and how they interact. The biologist, for example, may find it convenient to study an animal's nervous system, digestive system and circulation of the blood as separate subsystems. Each has its own properties and mechanisms and its own possibilities of malfunction, despite the close interdependence and high degree of interaction between them. Furthermore, each of the subsystems has a high degree of similarity between organisms, and even between species, which allows the subsystem to be developed as a specialized area of study. The zoologist, veterinary surgeon, doctor and dentist may each specialize in particular subsystems that form only a part of a complete animal.

A taxonomist may define subsystems within a community of organisms in terms of genetic affinities. An ecologist considering the same community would view the subsystems very differently, perhaps in terms of the position of animals in the food chain, or in terms of the ecological niches which they occupy. The different views of the same entities and their relationships can all contribute to the overall scientific understanding of the system. In geology, the reconstruction of the geological history of an area might lead to a system being defined as the deposition of the sedimentary rocks, with subsystems concerned with, say, the sediments and their movements in space and time; the agents, such as gravity, wind, rivers, tides and currents, which transported and deposited the sediments; and the assemblages of organisms, and their habitats, as revealed by the fossil occurrences. Events in each of the three subsystems take place through different kinds of process, and under different sets of controls. In a similar manner, the tectonic development of a set of folded rocks might be considered in terms of the input of patterns of stress and heat to the system, and their interaction with the geometry and viscosity distribution of the rocks to generate the pattern of observed strain.

Human limitations restrict the scope of any scientific investigation. It can be desirable to ensure that each system or subsystem has the simplest possible interface, or set of boundary conditions, with other systems. The difficulties of being compatible and consistent with work in other fields or other areas is thereby reduced. Within a subsystem, the same concept may apply, where for ease of description and explanation, the subsystem may be regarded as a set of compartments with comparatively simple attributes, each representing a state of the subsystem. The relationship of one compartment with another may or may not be understood. For instance, the relationship between two successive states of an atmospheric layer may be sufficiently understood in terms of the gas laws that equations can be derived to link the two compartments in the model. In other cases, such as the lateral transition between two sections of the same rock sequence, the underlying mechanism may not be understood. In such a case, the relationship can only be represented by a 'black box' which links the input and output, here the two compartments of the model, by an algorithm which is selected solely on the basis of the observed effect, without knowledge of the cause. Where the mechanism is understood, and simulated in the algorithm, the link is sometimes referred to as a 'white box'.

The use of simulation models in geology is described by Harbaugh and Bonham–Carter (1970). Examples of models that have some actual or potential relevance to geology may give an indication of some possible lines of approach. One model that is of considerable importance in geophysics predicts the value of a field, measured at a defined surface, resulting from the influence of a body for which the position, shape and relevant properties are known. Applications are in predicting the gravitational or magnetic field over a buried rock mass. If correct estimates have been made, then the field predicted by the model should be closely similar to the measured field. If the random elements are ignored, the relationships are sufficiently well defined that simulation is unnecessary and the position and shape of the rock body can be calculated directly from the observed data.

The effect at different levels of buried rock bodies has some resemblance to the three-dimensional framework of stratigraphic surfaces which was considered above at some length. The structural geology of the basement may exert an influence through an extensive series of deposits, as revealed in detailed isopach and facies maps, and can be considered as a field effect, possibly diminishing with time. The feedback between subsidence and rate of sedimentation may be positive or negative, depending on the sedimentation type. Positive feedback is shown by reef growth or the deposition of sand bars which takes place preferentially on relatively high areas of the basin floors, and may have the effect of emphasizing irregularities in the topography. Negative feedback may arise during a period of deposition of muds and channel sediments which result in overall smoothing of an irregular surface. Subsequent compaction may reinforce the pattern generated by deposition, if, say, muds that are thicker in the faster-subsiding areas also undergo greater compaction than the coarser-grained material. The effects of faulting, of folding of the deeper layers, and periods of uplift, erosion and unconformity add further complexities to the change of the underlying pattern through successive surfaces. It is not possible in these circumstances to deduce the underlying field effect by analytical methods, although simulation could provide a means of investigating the outcome of selected scenarios.

A number of other classes of simulation model can perhaps be mentioned, to indicate the role of such concepts in other aspects of science. The computer representation of such models is well established, and there may be scope for employing existing programs in a geological context. The transition, movement or change from one compartment of a system to another, gives rise to a variety of models. One is the filtering mechanism, in which preference is given to entities with certain attributes during the change of state. One geological example might be the transportation of sediment between a beach and the dunes behind the beach. Grains might be selected preferentially by size for wind transport. Subsequently, certain size ranges might be retained preferentially on the surface of the dune, depending on its geometry and the distribution of sizes already present. In a similar way, the geographical spread of chemical compounds from an ore body or hydrothermal deposit might depend on their mobility. The filter model is thus relevant where transport between regions is of interest, and the transport mechanism selects preferentially. An example outside geology is in information retrieval, where on the basis of a number of attributes, information can be filtered to be relevant to an increasingly specific set of enquiries.

Deposition or emplacement of sediment, igneous rocks or hydrothermal minerals is an indication that transportation into that region is more effective than the transportation out. Queuing theory is an aspect of mathematics which deals with the general problem of a set of entities arriving at an interface at random times. The interface has a limited ability to handle the input, and in consequence queues may develop. Examples of the applications of queuing theory are to lines of customers forming at the checkout points of a supermarket, telephone calls being queued at a switch-board which lacks the capacity to handle them all simultaneously, or information within a computer being held waiting in store, because, say, several programs required access to one line printer. The objective of queuing theory is generally to assist in the design of facilities by providing a means of exploring the consequences of a range of designs under various loads. It achieves this by giving a means of predicting the number, size and duration of queues, given appropriate parameters describing the pattern of input, the number and type of facilities for servicing the input, and the pattern of times taken for the servicing operation. If geological deposits are regarded as unserviced queues, therefore, a considerable body of mathematical work is available to simulate the process which caused this response.

The movement and spread of populations of organisms has been the subject of computer models in biology and ecology and an analogous approach has been taken to the development of fossil populations (see Craig and Oertel, 1966). Changes in the relative abundance of species through space and time is of interest to the paleontologist. The epidemic model, which, as the name implies, was developed in medical science, may be of interest here. The model considers the transfer of, say, myxomatosis through a rabbit population, spreading at a rate affected by the density of the population, and its susceptibility to the disease. The part played by agents carrying the disease, such as the rabbits' parasites, or the fleas of the black rat in the case of the plague, is also taken into account. The same mathematical model might be relevant to the spread of fossil species to environments in which they can find a suitable niche, and their colonization of the environments.

Study of the movement of entities between compartments, or transfer of material between environments or regions, can lead to a requirement to ensure that all the material is accounted for. The idea that the amount of clastic debris, in its various size classes, from the erosion of solid rock should correspond to the amount of clastic sediment produced from it leads to the concept of the sediment budget. Input–output analysis is used for analogous problems in econometrics (Leontief, 1966), where a matrix representation is used to show the various raw materials used by an economic unit, such as a country or an industry, and the amount of each raw material accounted for by each product or class of products. This representation clearly shows any discrepancies or omissions in the information, and indicates the potential consequences of a shortage of one raw material, or the implications of a change in output volumes. If the quantities of boulders, gravel, sand, silt and mud produced by erosion of the pre-existing rocks of a particular land area can be estimated, and estimates prepared of the quantities deposited in the various stratigraphic units formed from them, the Leontief input–output matrices would be a possible representation of that aspect of the process, see Fig. 74. Similar procedures could be followed in representing the allocation of chemical or mineral constituents to igneous rocks. Although in most geological instances, data are only very approximate, such methods provide a quantitative framework in which information from several sources can be assembled, and pattern, inconsistencies and gaps in the available information can be revealed.

Another interesting approach which can reveal gaps in a pattern is the set of growth models which have been developed to simulate the development of various kinds of organism (see Raup, 1966). A mathematical function can represent the sequence of shapes in the development of an organism, such as a mollusc, in which the various stages of growth are visible, as growth lines, in the adult. A change in the parameters of the mathematical function, can change the form of the simulated organism from, say, an involuted coil to a spiral. Some of the functions may appear complex, but far more complete information about the morphology of a species is presumably inherited by each generation through a genetic code. The simulation functions and their parameters provide a notation for representing morphological variations,

(a)

Input/Output	Boulders	Gravel	Sand	Silt	Mud	Totals
Schist	0·01	0·1	2·0	2·0	0·9	5·0
Granite	0·2	0·2	3·0	0·5	1·1	5·0
Paleozoic	0·0	0·0	0·5	0·3	1·2	2·0
Washed in from outside	0·0	0·0	0·0	0·2	2·8	3·0
Totals	0·2	0·3	5·5	3·0	6·0	15·0

(b)

Input/Output	Boulders	Gravel	Sand	Silt	Mud	Totals
Cretaceous	0·0	0·0	1·0	0·5	0·5	2·0
Tertiary	0·0	0·1	1·9	0·0	1·0	3·0
Removed from area	0·2	0·2	2·6	2·5	4·5	10·0
Totals	0·2	0·3	5·5	3·0	6·0	15·0

Fig. 74. Input–output matrices. In this example of a sediment budget, the output from one matrix determines the input to the other. Erosion of older rocks provides the output of (a). This is input to (b) to form the younger sedimentary rocks.

and a possible framework to consider the biological advantages and disadvantages of the various forms, including those which can be described geometrically, but are not known to occur in reality. In the same way, computer models can be constructed reflecting the surface morphology of eroded landscapes, or of surfaces of deposition under various conditions and environments. If they can be grouped and examined in a consistent manner, a comparison with the actual range of observed morphologies would be possible.

One class of models which strongly emphasizes the interfaces between compartments rather than the compartments themselves, is the class of finite-element models. Their primary application is in engineering studies, such as determining the stresses on the various parts of a bridge under different loads. The complex form of a bridge may mean that a direct analysis of stress at every point is impossible. However, the problem can be tackled by dividing the bridge conceptually into a number of elements of simple shape, for which stresses can be calculated, knowing the outside forces acting on that element. The method proceeds step by step, calculating the stress transmitted to each corner where finite elements meet, until stresses for the entire structure are defined. A similar procedure is used in reservoir modelling for studying the flow of hydrocarbons, or the flow of groundwater through an aquifer. The reservoir rock is divided into regions of regular shape which have reasonably uniform internal properties. The inflow of surface water into the appropriate regions, and the abstraction of water at wells and springs, are recorded in the model. Knowledge of the permeability, porosity and pressures provides the parameters for a model describing the flow of water through the system from one region to another, and allows the modeller to build up a picture of the evolution of the system through time, or to consider the consequences of various patterns and rates of extraction.

*4.19. A framework for the systematic use of expert opinion;
the risk analysis model*

Simulation and data analysis may throw light on various aspects of a geological problem. The systems approach should make it possible, having examined and defined the various aspects, to bring the threads together to meet the overall objectives in a consistent manner. An example of an objective with rather complex geological implications is the estimation of the reserves of hydrocarbons in a particular area. As it is an objective which is commercially important, the techniques have been highly developed within major oil companies, although some of the methods are not published in detail for commercial reasons (see Harbaugh et al., 1977). The approach is of interest in that it combines a variety of computer-based techniques. On the one hand, there is no means of arriving at a hydrocarbon resource estimate by deductive methods alone. The answer depends ultimately on judgment and experience. On the other hand, experts of high repute can arrive at answers which differ by a factor of ten. A high degree of accuracy cannot be hoped for, and estimates can best be expressed in probabilistic terms, reflecting the degree of risk inherent in accepting any given figure. As it can be anticipated that experts' views will differ, methods are relevant that have been developed for the systematic use of expert opinion, such as the Delphi method and the teleconference, described in the chapter on communication.

The crux of the systems approach to resource estimation is the analysis of the initial problem. The bald statement that the oil reserves of Ercu County are estimated at 250 million barrels does not contain much meat for scientific discussion. Nor is it obvious how a discrepancy with another estimate of, say, 300 million barrels, could be resolved, or even constructively discussed. The statement contains too many hidden assumptions for the geological implications to be clear, or for evidence to be examined to resolve the differing opinions. The first step, therefore, is to recognize that not all the resources are of the same kind. By considering them separately as 'plays' of more homogeneous character, the evidence relating to each can be examined in turn. One play might be the oil related to salt domes in the Permian, a second, oil trapped in anticlines in the folded Jurassic, a third, stratigraphic traps in Cretaceous sandstones, and a fourth, possible oil traps which might exist in the Devonian below an unconformity. For each play, the geological controls are different, and geological advice on each may have to be obtained from a different group of experts.

The second step is to consider which questions should be asked for each play. Some questions require a yes/no answer, such as: Is there a suitable source rock, and a suitable cap rock? Other questions require a quantitative answer, such as: How many reservoirs are likely to exist; what is their expected average area; thickness of pay zone; gas/oil ratio; etc? The analysis calls for a much more complex response than the original single question: What are the oil resources? However, each of the detailed questions refers to a specific aspect of the geology on which concrete evidence can be assembled and evaluated in arriving at a response. Where widely differing views are held, it should be possible in many cases to collect further evidence to resolve the disagreement. However, the various aspects are still matters of opinion rather than clearly established facts, and it is appropriate to frame answers in terms of probabilities which reflect the inevitable uncertainties.

The yes/no questions, like 'Is there a suitable source rock'? can be answered on a probability scale. A certain 'yes' would have a value of $1 \cdot 0$, a definite and certain 'no' a value of $0 \cdot 0$. Uncertainty, with either possibility regarded as equally likely, would be represented by a value of $0 \cdot 5$. This refers to the concept of probability described in §4.2. Where an answer cannot readily be given, it may be helpful to break down the question further, and list points such as geochemical evidence, knowledge of the depositional environment, and information on the same formations elsewhere, which could have a bearing on the answer. Weighting factors might even be assigned to each aspect. Quantitative answers, such as the estimated average thickness of pay zone, can be expressed as a probability curve, as in Fig. 67. The probability level represents degree of certainty, as described in §4.2, and reflects subjective judgment. However, the judgment concerns a specific geological feature which is clearly defined. Relevant evidence may be available from direct measurement, or analogy with similar situations elsewhere. For instance, questions about thickness of reservoir rocks might well be answered by an appropriate spatial model, in which observational data and hypotheses are already integrated, and possibly from simulation models.

While integration of many aspects of information may be a worth-while long-term objective, the time-scale is inevitably long, not least because of the time needed for geologists to learn and adapt to new methods. Furthermore, it is essential that data should not pass between two models, such as the spatial model and the resources model, without a careful and critical examination by a geologist who is completely familiar with the appropriate background. The information serves different purposes in the two models, and it cannot be assumed that two sets of data with the same name have the same meaning. As an example, a spatial model might have been constructed to show the most probable thickness of a porous sandstone at any point over an area. A contour map generated from that model might give a totally misleading picture of the likely volume of an oil reservoir, since it is not designed to indicate probable shapes of the sand body. The shape of the envelope of probable thicknesses might not even be a possible physical shape for the sandstone. At any rate, whatever supporting methods are used, the input to the resources model should be obtained by the judgment of one or more skilled geologists, and, though taking advantage of available supporting techniques, should not be delayed by waiting for spatial or simulation models to be developed.

The information from which the probability curve of Fig. 67 is generated can be in the form of three values: the smallest value that is thought to be a reasonable possibility, the most likely value, and the highest value that is thought to be a reasonable possibility. A triangular distribution can be calculated from these values, and is adequate for most purposes. Geologists who are more familiar with the method may estimate five points, perhaps the 0.05, 0.25, 0.50, 0.75 and 0.95 values, as shown in Fig. 75. With this information, the computer can take into account variations in the shape of the curve, and the geologist can express his views more accurately because

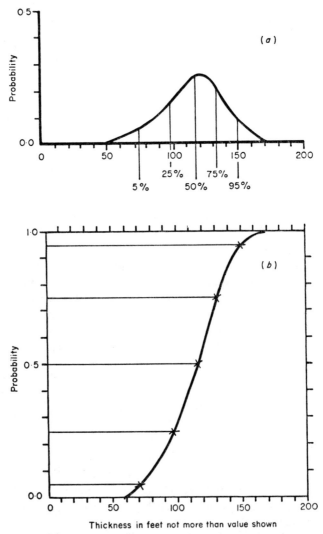

Fig. 75. Probabilities estimated at five points define the curve in (*a*) more precisely than in Fig. 67. The cumulative curve (*b*), showing the estimated probability of the thickness being less than the value shown, is a more convenient form of representing the same information.

less weight is placed on the extreme values, which tend to be particularly difficult to estimate. The concept of estimating probabilities is not an easy one, and although accuracy can only be achieved with experience, simulation techniques may help both in understanding the concepts, and calibrating the geologist's judgment. Consistency is important to the success of the resources model, and is difficult to achieve because not only are different circumstances involved in each play, but different groups of geologists, with different concepts and background, may be involved. The achievement of consistent estimates therefore requires that representatives from every group should take part in the discussions in which estimates are formulated. Disagreements occur within any group of experts, and an advantage of the probabilistic methods is that the uncertainty implied by the disagreement can be included in the estimate.

After probability curves and estimates have been agreed, the computer model is required to bring together estimates of the various parameters, and calculate the probability distribution of resources. Because the input data is in the form of probability distributions, this cannot readily be done by a single calculation. Instead, the results are obtained by 'Monte Carlo' methods in which the various probability distributions are sampled with appropriate random numbers, and the calculation repeated many times to give a distribution of results. The process is repeated until a sufficiently large sample of results has been generated to give a clear idea of the distribution. The distribution curve can be displayed as in Fig. 75(b). Perhaps the most important aspect of presenting the results in this way is that the inevitable uncertainty inherent in the estimate is clearly shown. The curve can be used in risk analysis. With fuel resources, obvious applications of the analysis are political and economic. A country or state dependent on income from the production of oil might decide to pursue a taxation policy which ensured adequate income at the lowest estimated level. An oil company, able to accept normal commercial risks, might decide to build a refinery which would be profitable if the 'most likely' value of the estimates was reached. A government decision to invest heavily in nuclear energy plants might be deferred on the grounds that if further exploration showed that the actual reserves were near the maximum probable value, the costs of the alternative energy source would not be justified. In each case, the risks can be quantified, and decisions reached on the basis of what are acceptable risks in any instance. This approach is particularly appropriate in economic geology, where quantitative estimates are directly relevant to the economic assessment of the prospect. But an analysis of the variables bearing on a problem, and quantitative assessment of a range of possibilities can also be valuable in a purely scientific context.

In summary, there are many computer models which can play a part in geological explanation. None of them supplants the human intuition and background knowledge where most explanations have their roots. But they do make it possible, by mathematical analysis and simulation, to test hypotheses and explore their consequences. More complex models can be constructed, of which the spatial model of the three-dimensional geology of an area is an example, in which data sets, known relationships, and hypothetical concepts can be brought together as a model in which information from several sources is combined to give a consistent pattern and improve the estimates made from the model. Integration of ideas at a higher level is provided by the risk analysis model, applied for example to resource estimation, in which information and estimates supplied from lower-level models or by the judgment of geologists is integrated in probabilistic terms within the computer. The last model in particular is closely tied to the communication of ideas. The role of the computer in this activity is considered in the next chapter.

5. COMMUNICATION

5.1. Communication in computer applications

Communication in the present context is concerned with transmission of information in geological applications of the computer. The need for transmission of information between data files and programs is self-evident, but not always satisfactorily achieved. Indeed, the ability to extract information from several data files, and to combine it for analysis has become the specialized task of data management. Similar requirements arise between computer models, although the solution is not necessarily the same. Overall, the transmission of information to be considered takes place among the following: geologists, systems analysts, programmers; support and maintenance staff and users of computer systems; computers, terminals; programs, data and models.

5.2. The shared coding scheme, standards and interfaces

There are a number of general concepts which are useful in considering communication, some of which have already been mentioned in connection with the recording of observations. They are the 'shared coding scheme', and, more familiar in the computer context, 'interfaces', 'standards', and 'protocols'. The shared coding scheme is a concept used in the study of human communication, where it refers to the subset of a language by which information can be accurately passed between individuals or groups. A shared coding scheme implies that the same vocabulary and syntax are known to both sides, that the context is known to both, and that words and phrases carry the same meaning. Precise definition and uniform usage are particularly important in the technical terms used in science, but, as variation in the usage of words like siltstone or greywacke suggests, they are not always achieved. There is also a risk that the same phrase carries different meanings for different groups. 'Core store', for instance, means different things to a geologist and to a computer scientist.

Within groups, and within subsystems, there is likely to be greater uniformity of usage than within the population, or the system, as a whole. For

227

example, a group of geologists mapping an area together, may develop an agreed understanding of the formation boundaries in the area without readily being able to communicate their definitions to an outsider. A group of programmers may write a set of programs and subroutines in which uniform conventions are used throughout. Variable names may be used in the same way in each program, data may be read in the same format, matching of argument lists may make it possible to call the same subroutines from several of the main programs. Each routine may be constructed in a similar way, with the result that any one of the group of programmers can readily understand the work of another. It is more difficult to link the programs to outside subroutines or data files collected by another group following another set of conventions.

The adoption of uniform conventions simplifies communication, and indeed is necessary at some level for communication to take place at all. Conventions which are fully defined and formally agreed are known as standards. They may be local; internal to a system, organization or discipline; national or international. The scope of the standards limits the area in which full communication is readily possible. However, the complexity of present day science makes full communication between all specialist activities impossible and probably undesirable. The emphasis in communication has therefore moved from definition of standards within systems to their definition at the boundaries at which systems interact, the so-called interfaces, and to the protocols, which are the set of rules and standards that are followed in communicating information. The selection of the interface between two systems involves determining the position of a boundary which may be somewhat arbitrary. Generally, more than two systems or subsystems are involved, and it is important to define the interfaces to allow flexibility of interaction. Selection of the interfaces between subsystems, of course, also determines the scope of each subsystem.

A concrete example may make the significance clearer. In the last chapter, the assessment of hydrocarbon resources was seen as an activity which could involve a number of computer models, and several groups of geologists, all contributing knowledge from their own background knowledge, experience and assumptions. Good communication between them is essential. Communication between computers, and from computers to remote locations, are also likely to be required. Access to external sources of information, such as bibliographical data bases and special-purpose programs or devices may be desirable, and the end product of the investigation may well be information which is needed in computer-readable form for input to another external program, perhaps for economic evaluation. The example, though complex, is by no means unrealistic.

The systems analyst must view the entire activity as a complete system with defined objectives, namely, the assessment of the hydrocarbon reserves. It contains several partly autonomous subsystems, and is interfaced to several more. For the most part, these are outside the control of the systems analyst, who is thus constrained to work with already defined protocols. But there are other areas in which he may have complete freedom to recommend the scope and boundaries of the subsystems. Within a geological organization, it is likely that there would be freedom to construct the three computer models already mentioned: the data base, the spatial model and the resources model. As is often the case, the decision on whether to regard these as three autonomous systems or one single integrated system can be seen in terms of a trade-off. The greater the degree of integration, the greater the degree of consultation and agreement necessary in defining standards. An administrative mechanism might be needed to ensure that the standards were in fact followed, and that where an element of ambiguity existed, it was not resolved unilaterally by a single group. Familiar procedures which work well in an existing framework might have to be abandoned in order to meet new standards. If, on the other hand, integration is not attempted, there will inevitably be duplication of effort in designing and writing programs, and ad hoc measures will be necessary in transferring data between systems. A piecemeal system will also be difficult to maintain and to document, and the absence or departure of one individual could cause a gap in the flow of information which would be difficult to bridge. From the user's point of view, the diversity of procedures may make the overall system unnecessarily complicated.

The extremes of bureaucratic control and anarchic confusion are both undesirable, but careful study of the system may suggest that efficiency and ease of use can be achieved without loss of local control. The arguments are parallel to those for centralization or decentralization of any activity. The complexity of the factors that have to be taken into account means that no perfect solution is likely to be found, and constant minor changes of plan may be needed to accommodate individual personalities and different rates of progress in different groups. The overall strategy must therefore be sufficiently flexible to cope with such changes.

5.3. Information flow and the system of geological communication

A starting point is a consideration of the information flow in existing systems. A crucial point is the number of individuals who are able to communicate regularly and routinely to the extent necessary to maintain a uniform and consistent approach. The number depends on the complexity of the topic, the organization of the group, and the means of communication. It is unlikely to exceed twenty and could be very much less. One person may be a member, formally or informally, of more than one group, probably not adopting the same role in each. The 'technological gatekeeper' role, the individual who specializes in bringing new ideas to the group with whom he works, is particularly important where major innovation, such as the introduction of computer systems, is being attempted. The gatekeeper, in all probability self-appointed, is particularly likely to be a member of several groups, perhaps a geochemist who makes a point of working closely with a computer group, so that he comes to be regarded as a 'computer expert' among the geochemists with whom he works.

The main mode of communication within a group may be personal contact, but larger, less tightly knit or more widely dispersed groups may communicate primarily by conferences, committees or through papers in the scientific journals. These mechanisms provide a means of communication within a much larger body of scientists, who are peripheral members of the group, not concerned with sharing information on a day-to-day basis. The level of detail and the frequency with which information is shared varies greatly, from the scientist who works alone for three years and publishes the results in one short paper, to the computer operations manager who consults the shift diary and discusses progress with the operators once an hour. Consideration of the mode and frequency of communication, and the nature of the information being transmitted, may lead to an identification of groups concerned with particular aspects of information and the pattern of information flow among them.

The emphasis in analyzing the system as a whole can then be placed on the relationships between subsystems, and communication between them, rather than on the detailed processes within each subsystem. The entire system is thus considered in modular form, with the overall description no more complex than that of a subsystem. Efficient design and ease of use of the overall system can be achieved by establishing and maintaining standards for the interfaces between subsystems. It is particularly important that where one system interacts with several others, such as the interface between the user and several models, that one consistent set of standards should be maintained. Otherwise the user is put in the unfortunate position of having to learn a new set of rules for instructing the computer with every model he uses. Once interface standards are agreed, the designer of the subsystem is left with considerable flexibility in the procedures for meeting them. For internal efficiency, or because of pre-existing conventions, he may choose to have a translation module between the input or output standards of the subsystems and the interface standard. An example of this approach was given in §2.13 in terms of formats for storage, processing and exchange of data.

5.4. Communicating with the computer; interactive and batch processing

Returning to the main example, an attempt to identify interfaces between subsystems might give the results shown in Fig. 76. The physical aspects of data communication are the most firmly established and can be considered first. An important aspect of communication between the user and the computer is the expected response time between submitting a task to the

```
                                A            B           C         D
A  Hardware                 1 2 3 4 5 6 7 8  1 2 3 4 5 6 7 8  1 2 3 4 5 6  1 2 3
   1 Terminals        ..   ..
   2 Modems ..     ..    .. X
   3 Multiplexor    ..   .. X
   4 Communications processor X
   5 Mainframe computer 1 ..     X
   6 Mainframe computer 2 ..  X
   7 Leased telephone link  .. X
   8 Dial-up telephone link .. X


B  Software and data    ..
   1 Operating system     ..    XX
   2 Communications software    X
   3 Executive program   ..     X       XX
   4 Applications programs      XX      X X
   5 Data management system     X       X ?X ?
   6 Bibliographical DBMS ..     X       X   ?
   7 Geological data base  ..    X       X     ?X
   8 Bibliographical data base   X       X       X


C  Staff
   1 Geologists     ..    .. X      ? ?      ?XX ? ?XX
   2 Systems analysts    .. XXXXXXXX  XXXXXXXX  X
   3 Information scientists .. X    ?X      ? ?XX ?X ?X  XX
   4 Applications programmers X   ?XX    XXXX ? ?X ?   ?XX
   5 Systems programmers  .. XXXXXXXX XX ? ?XX ? ?   X ? ?
   6 Operations staff   .. XXXXXXXX XX ?          ?X ?XX


D  Organizations and groups
   1 User departments    ..                          XXX ?
   2 Standards forums    ..      XX       XXXXXXXX  XXXXX    X
   3 Computer user groups .. XXXXXXXX  XXX  XX     ?XXXXX  XX
                            1 2 3 4 5 6 7 8  1 2 3 4 5 6 7 8  1 2 3 4 5 6  1 2 3
                                A            B           C         D
```

Fig. 76. Interfaces in a system. The components of a system are listed in the rows of a matrix, and repeated as column headings. A significant interface between two components is shown by an X where the row and column meet, and a possibly significant interface is shown by a ?.

P

computer and receiving the result. This varies from a fraction of a second to several days. With the longer response times, a complete job is submitted in batch mode. That is, the job is entered in a queue of jobs waiting to be processed on that computer. Depending on the priority assigned by the user, and the availability of the resources required by the job, such as memory, disk, line printer, magnetic tapes, program and data files, the operating system schedules the job to be run as soon as it can be fitted in. The operating system is a complicated program in its own right, and is a significant indirect cost, even on a large computer. It ensures, however, that efficient use is made of the rest of the machine's resources. The time taken to receive results depends on how heavily the computer is loaded and on the method of submitting jobs and returning the output. On some large computers, a 'cafeteria' system, restricted to small, high-priority jobs, is operated at certain times of the day. The user places his job on cards in a card reader, and by the time he has walked to the line printer, his output may be ready for collection. The turn-around time, that is the response time for a batch job, is often two to three hours at a lower priority or on a more heavily loaded machine. The job is submitted over the counter to the machine operator, and the results are returned in the same way. Very large jobs, or those requiring special facilities, may be held for an overnight run, or even for weekend running, when they interfere less with the normal operation of the computer. A slow response, of a week or more, results from jobs being submitted and returned by mail from a distant computer. The use of remote job entry to reduce this time is discussed below.

The normal development of programs and analysis of data requires a reasonably short turn-around time. This is because these processes tend to require repeated runs, either for jobs which failed because of an error, or analyses which must proceed on a step by step basis. If turn-around is too slow, the user is unable to give the work continuous attention over a period, and much time is wasted trying to remember the previous stage. The response time of the user to the machine, however, should be long enough for careful consideration of results before the next job is submitted. A user who feels he has to work at the pace of the machine may waste time through undue haste. For this reason, fast turn-around may be more necessary on simple than on complex tasks.

Interactive, as opposed to batch, processing involves the user in typing instructions to the computer at a terminal, where he should receive a rapid response. This implies that each instruction is comparatively simple, and if the response takes more than a very few seconds, the user wastes an unacceptably large proportion of his time waiting for the computer. Each instruction may be no more than a line in a program. Many computer languages, including FORTRAN, have been modified to provide interactive versions, but languages like BASIC and APL are more suitable, having been designed specifically for interactive use. Complete subroutines and programs can be invoked interactively. Utility programs for editing and amending data and program files are also available on interactive systems.

5.5. Terminals, data links and networks

The need for interaction arises when processing is a step by step operation, as in data analysis, retrieval, and some simple modelling, or when it consists of a sequence of small operations, the results of which should be checked immediately, as in data entry or editing. It makes possible a dialog between the user and the program, in which the future course of the analysis may not be known at any point, but is dependent on the results obtained up to that time. In interactive working, the user normally operates from a small terminal with keyboard and slow (say, 30 characters per second) printer or VDU on which the dialog is displayed. It is also possible on some computers to interact with graphic data displayed on a screen, using a small fixed set of basic instructions, or 'menu', for such operations as change of scale, shift of origin, delete, move, insert line, etc. The operation can be selected by pointing to the appropriate box on the screen with a light pen, and the particular line or point on the diagram on the screen, again with the light pen or similar device.

These activities need not take place in the computer room, and communications facilities make it possible for the geologist to work from a terminal in his own office. Access to the computer from distant points is made possible by 'modulating' the binary signals used by the computer to a form suitable for transmission along a telephone line, and 'demodulating' the signal to an appropriate binary form at the receiving end. The modulator–demodulator unit or modem performs this function at each end of the transmission line, thus translating from the standard in the computer or terminal device to the standard used for transmission. The standards within the telephone network are set by the PTT's (the Postal, Telephone and Telegraphs organizations) internationally through two collaborating bodies, the CCITT (which refers to the International Telegraph and Telephone Consultative Committee) in Europe and EIA (Electronic Industries Association) in America. They specify the voltage levels and other characteristics of the signal, normally, of course, representing speech rather than data, which is transmitted on the telephone system. The modems at each end of the transmission line have matching characteristics, and the signal entering one of the pair is thus essentially the same as that leaving the other. As far as the computer is concerned, the effect, apart from some possible loss of signal quality due to line noise, is that there is no apparent difference between a terminal in the same room, and one connected by telephone line from another town, or another continent. Where line quality is adequate, the modem, which is physically connected to the telephone line, can be replaced by an acoustic coupler. The coupler is attached or built into the terminal, and transmits and receives by sound through the normal telephone handset. It is suitable, therefore, for portable terminals, enabling them to be used from any telephone.

Although the difference between a distant and a local terminal may not be apparent to the computer, it can be rather apparent to the user in the form of telephone charges. Where many users communicate between the same two points, a device known as a multiplexor makes it possible for several terminals to transmit simultaneously on one line. Alternatively local storage devices such as floppy disks or cassette tape can store messages entered from the keyboard, and transmit the complete file at a faster speed. A minicomputer can also act as a message concentrator, handling input from a number of slow terminals, and transmitting batches of information, as appropriate, to a remote computer. If there is heavy communication traffic, another possibility is remote job entry, in which relatively fast devices, such as a card reader and line printer on a local mini-computer transmit complete computer jobs to a large remote computer, and return the output resulting from them. The remote job entry system operates in batch mode, and in many applications is more economical than the slow terminal which operates in interactive mode.

Protocols again apply to the interfaces between the terminal and the computer. The remote job entry station, or remote batch terminal, as it is sometimes called, submits a complete job, and therefore interacts with the operating system of the computer. Although protocols such as HASP and HDLC are widely used, the protocol tends to differ between manufacturers, and even between operating systems. The remote batch data is transmitted synchronously, which means that the transmitted and the received signals are kept in step by timing pulses. Typical transmission speeds are 2400, 4800 or 9600 bits per second (bps), which can cope with devices of approximately 240, 480 or 960 records (cards or lines) per minute for input, output or both together. A system able to transmit input for one job and simultaneously receive output from another is said to operate in full duplex mode, as opposed to half duplex where transmission is in only one direction at a time.

The expense of dialled telephone calls can be reduced, where there is heavy information traffic for long periods each day, by leasing lines permanently from the PTT. The lines can be kept open when not in use, thus avoiding the need to re-establish connection by dialling. There are other methods of transmitting information between computers such as short-wave radio, telegraph and television cables, but the telephone network, because it offers a communications system world-wide, for which the capital cost has already been incurred for speech facilities, carries the main bulk of computer communication.

The slow terminal, operating typically at 110 or 300 bps, equivalent to 10 or 30 characters per second, transmits in asynchronous mode. This means that each character is transmitted as a separate entity, a sequence of eight bits with the beginning and end flagged by a start and stop bit. The character codes for asynchronous transmission are a widely accepted standard, and a wide range of computers can thus be accessed from most slow terminals. The same protocol has been adopted for keyboard and printer devices, visual-display units, and devices with local storage in the form of cassettes tapes, floppy disks or paper tape. Digitizers and many other specialist devices can be provided with the same interface.

Although many types of computer can accept the characters from a slow terminal, there are marked differences in the commands and the control and editing conventions and languages in different computers. As a result, although physical communication is straightforward, the user may have to understand many computer systems in order to take advantage of the flexibility. At this higher level of communication, there is little general agreement on protocols. This is something of a barrier to communication, and the next point to consider is the reasons why the flexibility of communicating with many computers and between computers may make it desirable to overcome that barrier.

When communications link several computers, the result is known as a network. Perhaps the simplest form of network is the so-called 'star' network, in which a number of sites are all linked to one central point (see Fig. 77). The communications protocols are relatively easy to define, since the interface between a remote device and the central computer is the only one that need be considered. It may be possible in a star network to communicate between two peripheral sites by passing the information to the central computer as one operation and from there to the second peripheral site as a second operation. If direct interactive communication is required between the peripheral sites, then the points in the network must be connected in a more complex manner, and the number of interfaces becomes very large. Standard protocols are then needed throughout the network to ensure that any two terminals can communicate.

The aim of creating a network is generally to share facilities of various kinds; to avoid duplicating resources; to give alternative fall-back facilities when equipment fails; and to enable various sites to access or contribute to the same data bases or program libraries. The network may also give advantages of scale, by sharing larger and more economical computers than any single site could afford. Development plans can be more flexible and the

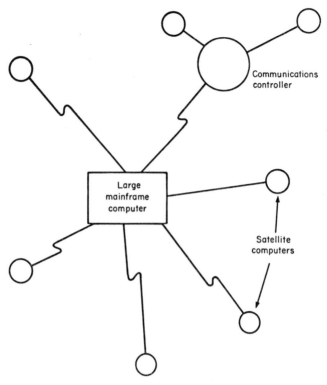

Fig. 77. A star network. Seven small computers are linked to a large central computer, two of them through a communications controller.

periodic under- and over-loading of equipment can be reduced by spreading the load among a number of machines or to wherever the capacity is available to handle it. Users of a network have a wider range of facilities at their disposal than would otherwise be possible, and communications lines, terminals and work stations can be operated in a more cost-effective manner because they are more widely shared. The so-called distributed processing which networks make possible gives greater flexibility in siting data bases, package support and other resources. Finally, access to a network allows a greater degree of specialization at individual sites, with a possible major saving in cost. For instance, a multi-user minicomputer for data entry and interactive processing of currently active files may be an adequate, and much cheaper substitute for a general purpose computer if good access through a network is available for running large jobs elsewhere and for access to the data archives and models held on a large remote computer.

The development of networks brings into clearer focus the various problems which arise whenever information is communicated on a large scale. Comprehensive network standards have been defined at a physical level which cover protocols for data transmission and file transfer. They also embody administrative mechanisms to ensure that the published standards are not amended, nor ambiguities resolved in an arbitrary manner. Transfer of files between computers or computer systems is commonly on magnetic tape and, although several standards exist, installations generally have enough flexibility that matching standards can be found. The data within the file may have to be rearranged and altered to transfer them from one software environment to another. Bridging software to perform the necessary amendments requires knowledge of the job control language at both installations, and may be most readily prepared by the two groups working to an agreed exchange format. Although magnetic tape is suitable for occasional exchange of large data files, smaller data sets can be transmitted directly where data links exist. Network conventions define the 'frame' within which the file is transferred. This might consist of flags to indicate the beginning and end of the frame, header records identifying the file and indicating its destination, and a trailer record, immediately before the final flag, giving a checksum value to detect transmission errors, calculated as the data are transmitted, and compared with the value calculated by the same algorithm when the data are received. In a complex network, there is a risk of data being accidentally transmitted to the wrong recipient. With confidential data, information within the frame can be scrambled by an encryption algorithm, so that it is meaningless until unscrambled at the receiving computer by the corresponding algorithm with the same parameters.

The frame in which the information is contained serves a similar purpose to the envelope that contains a letter carried by mail. It ensures that the contents are kept together, that they go to the right destination and can be identified on arrival, and that, if the seal is unbroken, all the information which was sent is actually received. The information in the frame can be of various kinds, such as a program, a set of data relevant to a particular model, or a set of plotter instructions for drawing a contour map. In order to convey meaning to the recipient, the information must fall within a coding scheme shared by the originator and the user. If a link is used only for, say, graphic data, then the shared coding scheme is not extensive, and one data set can follow another without additional explanation. Where the shared coding scheme is larger, descriptive material must accompany the information to identify its type and characteristics. The descriptive material accompanying the information is part of its documentation, which has an important role in communication, whether transmission is through an on-line computer network, or as magnetic tapes or coding forms.

5.6. Documentation; exchange of programs and data

Documentation is generally regarded as an unwelcome chore by the systems analyst and programmer, and can be kept to a minimum by making as much use as possible of existing documentation, by codifying routine documentation, and by resisting the urge to document for its own sake. Some information will never be used more than once, and some is cheaper to recreate than to store. It should be discarded, not documented. In general terms, the function of documentation is to use one shared coding scheme, such as English, to expand a more limited coding scheme, such as the computer data record. Three levels of coding scheme are commonly apparent to the geologist in computer data. The first level, the data file itself, has direct meaning to the group who collected and used the data with their computer programs. The second level, which is not always present, is an encoded data description which enables a larger group to understand the content of the data file. The data description may contain information required by the computer for processing the data, such as the format and arrangement, as well as describing the data content in outline for the benefit of the user. The form of the data description is rigidly defined, and is generally restricted to a single system. The schema and subschema, described in §2.16, are at this level.

The third level of documentation is a report designed to be comprehensible to anyone with the necessary technical knowledge. Its purpose is to provide a description which enables other scientists to understand and utilize the data. This third level employs a widely shared coding scheme, such as technical English, to explain the background of the geology, sampling scheme, etc., and may take the form of an internal report or a published paper.

The information which is exchanged may be programs rather than data, in which case the use of a widely accepted programming language makes communication easier. Documentation for the language presumably already exists and need not be repeated. The task is made more laborious, however, if non-standard features of the languages are used, since these require clear identification, probably in comment statements in the program, and separate documentation. The majority of programs in a scientific environment are used a surprisingly small number of times for production purposes, and it is by no means unusual for a program to be run more frequently during development and testing, than in actual production of results. In these circumstances, there may be no need to communicate information about the program, other than the mathematical method used to obtain results. The best strategy may be to use a programming language which is convenient for program development, probably an interactive language such as APL, without taking time for documentation. The completed program cannot readily be made available to others, and it may even be questionable whether such a program should be kept rather than rewritten if it should be needed again. At the other extreme, some programs will have frequent use by many users over a long period. Examples are the operating systems, utility programs and compilers generally provided by the computer manufacturer. More directly related to geological applications are packages for statistical analysis, simulation, graphic presentation, spatial analysis, contouring and data management. These areas each involve complex programming systems which could represent 30 or more man-years of effort in designing, programming, testing and documenting the system. One reason for the size of the task is that they are designed for use on a widespread basis, and must therefore be fully documented and rigorously tested. The ability to transfer the programs to a new group of users is an aspect of communication which places a major overhead on the original development. The programs must, if possible, be written in a widely used programming language, using only the standard features, to ensure that the shared coding scheme extends across a wide range of installations. The source version of the programs may or may not be made accessible to the user, since by deliberately restricting communication of this aspect, the commercial value of the package can be more readily maintained.

Licensing agreements may be entered into by the supplier and the user, again for commercial reasons. For internal purposes, the software must be fully documented, since a number of programmers are inevitably involved in such a large task. Correcting the errors which occur in all software, or continuing the work of a programmer who leaves, is dependent on good documentation. Guides to suitable procedures are available, see NCC (1970), and can be followed with advantage in any large software development. Even on smaller jobs, care should be taken in writing a program to ensure that comments are included among the program statements to clarify the function of each section of the program, even if this seems self-evident at the time. An information sheet or report accompanying the program should state clearly what the program does, how it does it, what options there are in the program, what parameters are supplied by the user, and what error messages it produces. It should always include the date, the programmer's name, and the model of computer on which it was tested. An example of input and output from a test data set is helpful.

Another area of documentation for a large software package is the user manual. This tends to be rather widely distributed, and may be seen as a means of helping to sell the package. The information in a user manual corresponds to that in the information sheet or report for the smaller program. It may include instructions to the user on how to use the program, how to specify the selected options, the form in which the data should be supplied, and an indication of the form and meaning of any error messages that may be produced. It should preferably include examples of input and output. Another aspect of communication which is important in a widely used package is that the output should be clearly labelled, and as far as possible self-explanatory. Any necessary additional explanation should be set out clearly in the user manual. Finally, an advertizing brochure may be needed, setting out the purpose and advantages of the program succinctly, to enable a prospective user to determine whether or not the program is of potential interest to him.

The majority of geological programs fall between the two extremes of the program which is used once and discarded, and the program which is widely distributed to many users. In the first extreme, the cost of the program is borne entirely by one application. In the other case, the costs may be spread over many users, but a major additional overhead is incurred for communication. Indeed, a major software package requires a permanent support and development group to advise users, and maintain the programs as hardware is changed, new techniques are developed, and errors come to light in the original program. These costs are seldom justified except within a large organization, or by widespread external use. Marketing of software is itself a specialized activity, although it may be possible to arrange this through a software house. An alternative is publication, particularly if the program was developed for academic purposes. The basic ideas can be published in an appropriate periodical, and the user is expected to do his own work in modifying the code to work in his own circumstances, and may have to correspond with the author to obtain details which are not published.

From the point of view of the user, the programming strategy that he follows will obviously depend on the circumstances. The first point to consider is whether existing programs can do the job. The cost of a commercial package can be high and is justified only by heavy usage. But it may be possible to use the package through a computer bureau or other large external organization where the software is available as part of the service. A considerable amount of software is available without charge from computer manufacturers and elsewhere. The ephemeral computer literature provides a guide to possible sources. Publications are another possible source of programs, where direct communication with the author can sometimes be rewarding. The user must then be prepared to do some work on the program himself and should be familiar with the language concerned. For small programs and for highly specialized tasks, it is usually desirable to write the programs in-house.

5.7. Communication between programs

In a complex programming application, the user may find that he is using programs of several types from several sources for the same or related data. He is then faced with another problem: how to communicate between programs. A set of programs in different languages, each written in a personal style, requiring data, options and formats in different formats, presenting the output in different ways, with some functions overlapping and some totally missing, is a user's nightmare. It is therefore important, as with any other aspect of system design, to plan program acquisition carefully. The available computer equipment and the associated systems software and communications links have an important bearing on the design. The processing system may fall into a number of applications subsystems, like the data base, the spatial model and the resource model mentioned above; or it may fall into a number of activities related to hardware components, such as on-line data entry, interactive editing and data validation on a local minicomputer, and spatial analysis on a remote batch computer. Having defined the subsystems, a high degree of uniformity should be sought within each, and interfaces designed to give as simple a route as possible for data that has to be transmitted between them. The user interface is particularly important, since confusion and wasted effort will result from different sets of commands within each subsystem.

An example of this difficulty is provided by library enquiry services, which in some countries have access to a number of bibliographical data bases, each run by a different organization with its own conventions, thesauri, access languages, etc. One solution that has been adopted is to develop a set of commands which is used by the library and is translated by a small local computer to the appropriate conventions for whatever data base is being accessed. With geological data, it is probably more usual to acquire the necessary data from outside sources and modify and reformat them for the local computer. The advantages of a communications format and the translation process were mentioned above. A similar approach can be taken in linking models in one organization. If necessary, a controller program (see §1.30) can be written to offer a common language to the user, and to generate programs for the various systems. This is seldom necessary, however, if the subsystem boundaries are well chosen, since most users will require access to only one subsystem. A reasonably comprehensive software system within a geological organization could probably be based on no more than three packages, and additions, amendments and bridging software can be written following the same conventions as the main packages. The command structure and conventions are then those which were thought by their designers to be most suitable for the applications served by the package. Users of large mainframe computers may find that documented software for their principal requirements can be obtained, complete with links between the programs, from the manufacturer or bureau.

5.8. Communication between the geologist, systems analyst and programmer

Following this approach, many requirements can be met without much in-house documentation of programs. The standard of in-house documentation does, however, need some careful consideration. The extent of future usage will determine whether or not a high standard is necessary. For the majority of small-scale applications, the geologist who knows a programming language is able to write and test his own program. For development of larger systems, or where the geologist is not familiar with computer methods, the systems design and the programming activities may be handled separately by different groups. Another communications problem then arises. The systems analyst should preferably be familiar with the programming language used, and the programmers should know something of the system, since an overlap of knowledge is a prerequisite for communication. The formal documentation of a system should be brought to a level where it gives good guidance for the programmers on a day-to-day basis. Pro forma documents can be obtained for this purpose (NCC, 1970), and probably already exist within an organization sufficiently large to justify this approach.

A more common communications problem exists for the geologist who wishes to have new programs developed, although he does not himself know a programming language. In fact, a geologist in this position is probably unfamiliar with existing programs, and might well find that an already existing program could adequately meet his needs. A geologist who intends to use the computer extensively should find time for the few days needed to learn simple programming, if only to enable him to match the geological requirement with a knowledge of possible computer solutions. Even the geologist who can write a program, however, may need help from a professional programmer in preparing a program which is efficient, easy to use, and easy to understand.

5.9. Word and text processing; the teleconference; the Delphi

Communications between individuals, in a broader sense, takes place conventionally by discussion, committees, conferences, telephone, telegram, telex, facsimile transmission, by memorandum and letter, reports and informal documents, microfilm and microfiche, and published reports, papers, maps, diagrams, abstracts, indexes, review papers, books, dictionaries, directories, encyclopaedias, etc. The conventional methods are of interest in computer applications, first, because any application of the computer has to fit in with the conventional methods, and secondly, because computer methods are beginning to impinge on the more traditional procedures. The methods of communication listed above differ in the size and character of the communicating group, on whether the record is ephemeral or lasting, of immediate or long-term significance, and whether it is in detailed or summary form. As indicated above, the same considerations apply to information in a computer system.

In addition to the applications which involve data specifically prepared for computer processing, and the programs and models which control the processing of the data, there is an important area of computer applications in word and text processing, which is concerned with information that may be selectively retrieved and rearranged, but is not otherwise significantly modified by the computer. The information on input has direct meaning for the user, and retains the same meaning on output. An example of a simple word-processing device is an electric typewriter with a memory which enables it to store a sequence of characters for editing or re-entry, if need be, before typing. It has similarities to the computer terminal, from which an entire file can be entered, stored, edited and retyped. From a computer terminal, pre-existing documents or files, or information selected from them, can be selected and retrieved, rearranged and combined, and typed. Depending on the extent of the network to which the terminal is connected, the output document can be directed to another user at another terminal, or to a specialized output device, such as a line printer, a microfilm camera or a typesetter. The computer network can thus give considerable flexibility, not only in assembling the text as a document, but also in preparing copies at distant points and in various alternative forms, from output on a VDU for quick scanning to microfilm and camera-ready copy for publication.

The flexibility which this provides has led to the development of the teleconference (see Vallee and Wilson, 1976), in which a framework is provided for several participants at a number of locations to contribute their views on a defined topic. These can be added to or questioned by the other participants, just as ideas are expressed and extended or refuted at a con-ventional conference. An important difference is that the record of all entries to the teleconference are available to each participant on his terminal at any time. Thus, while a fixed time may be allotted for questions after a talk at a conventional conference, the comments at a teleconference can be raised in any order and at any time. Teleconferencing procedures can be combined with the Delphi method introduced in the last chapter.

Delphi may be characterized as a method for structuring a group com-munication process so that the process is effective in allowing a group of individuals, as a whole, to deal with a complex problem (see Linstone and Turoff, 1975). It is most often concerned with the estimation of quantitative values, possibly requiring an assessment of probabilities as described in §4·19. In experiments, it has been shown that a combined estimate, based on the opinions of several experts, is more likely to be correct than a single view. Since the Delphi method is relevant only when direct factual evidence is not available, and hence conclusions must be based on expert opinion rather than observations or data, the results can be proved correct or otherwise only on an experimental basis, where information is deliberately withheld from the participants. A Delphi usually has two or more rounds, in the first of which opinions are obtained and summarized. The summary is then studied by the participants, and a second round of opinions is obtained and summarized. It has been found that the second summary is sharper, and more nearly correct (in experiments) than the first. The objective is not to force a consensus, and the Delphi should not conceal disagreement, which can readily be expressed by probability methods (§4.19).

Experts may have different levels of expertise, both overall and in specialized subject areas. It is possible to allow for this within the Delphi by requesting the participants to assess the level of their own knowledge, and the value of their own opinion, on a scale from $0 \cdot 0$ to $1 \cdot 0$. Their responses can be weighted accordingly. Again, experiments have shown that self-assessment can improve the result. Participants may be asked to take part anonymously, as this reduces the risk than an opinion will be given too much or too little weight because of the forceful or the negative personality of the protagonist. Experiments, some of which are quoted in Linstone and Turoff (1975), have indicated that these features can improve the communication process, and thus make it possible to arrive at a more accurate result. However, there is an artificial element in any such experiment which may make the conclusions invalid in other cultures and circumstances. Delphi offers a battery of techniques from which can be selected the weapons that best meet the needs and traditions of the group concerned. While the method may be useful in many fields, such as obtaining views on future requirements for computer facilities, by far the most important geological applications to date have been in the preparation of resource estimates.

5.10. Information systems; catalogs and indexes

More mundane applications of computer text processing are in supplementing the typing pool and replacing some of the minutes and memoranda which are part of the daily flow of information in any large organization. The flexibility of output also assists in the preparation of catalogues of specimens in a museum collection or books in a library, where the computer is used to rearrange and abstract the information in various ways, and assemble it in a suitable form for publication as a set of catalogues. A data base may be prepared at the same time for direct computer retrieval. With libraries, potentially including data libraries, bibliographical information tends to be disseminated first for 'current-awareness', probably a list of recent acquisitions, which can be scanned to look for current material of particular relevance. The current-awareness information may then be included in the main data base for routine catalogue production and for selective dissemination of information (SDI). The purpose of SDI is to retrieve data which matches a 'profile', that is, a list of codes or keywords which defines the interests of the user. The information supplied by SDI should thus be more directly relevant to the user than that supplied by a current-awareness service. A data base built up over a period of time can be accessed for retrospective searches in which a user wishes to retrieve information that matches a particular profile.

The various cataloguing and data retrieval activities mentioned above generally provide output in conventional printed form and the user may not even be aware that a computer was used. The computer-produced index has the purpose of directing the user to the source of information, rather than of supplying the information directly. The index may be generated by searching for defined keywords, or may be based on descriptors or codes supplied by the indexer. Indexing information may prove to be one of the most useful areas in which to begin the computerization of large data files, possibly giving the best return for comparatively little effort (see Farmer and Read, 1976), and may indicate whether or not further effort on computer input for that data file would be worthwhile. By leading the user to the sources of information, the index is another important tool in communication.

In summary, communications are an increasingly important part of computer applications. A major reason for bringing general, quantitative, mathematical concepts into geology was the need to share data and methods in a larger group. The computer can readily handle data in this form. Many geologists can share the effort and the benefits of writing and maintaining programs, and can contribute and use data in a shared data base.

The digital representation of data is as appropriate in communication as in processing. User terminals and remote batch terminals can be connected to a central computer through data transmission systems, and computers can themselves be linked in complex networks, to share facilities and re-sources, and provide the individual user with a wider choice. Word and text processing systems make additional information available in computer-readable form from which can be extracted items for data retrieval, processing and display, as well as leading to new modes of communication, such as the teleconference and Delphi methods. A large part of all geological information that is now being recorded will be in digital form at some stage, and could be available to the computer. There is thus a need to design systems which will ensure that data can be made available in an appropriate form where and when they are needed. A description of information flow, the implementation of standards, and a careful examination of interfaces and the communication of information across them are essential aspects of the systems analysis and design.

6. IMPLEMENTATION

6.1. Defining the objectives

A successful application of the computer should meet the geologist's objectives, and take full advantage of the computer methods that could be of benefit. Since the geologist must play a central role in any geological application of the computer, the question of implementation is considered here from his point of view, looking at what he should contribute to a multidisciplinary activity, what he can expect from the computer specialists, and what is involved, if, as is not unusual, the geologist is also the systems analyst and the programmer. It is necessary throughout the process of systems analysis and planning to keep in mind aspects which have a bearing on the practical outcome, and to consider points which help to find the best method of implementing the conclusions. The ideas, after all, are of value only when they are put into effect. Therefore, it may be useful to summarize, in this chapter, a number of points, most of which have been mentioned earlier.

To an increasing extent, geologists are working within systems where the design work is already complete. In large systems, most of those involved are responsible only for a small part of the total system, and their own contribution must be made within a framework of decisions reached at an earlier stage or in a wider context. As the subject matures and the overall framework is better defined, it becomes increasingly important to understand and respect what already exists. Continuity of effort counts for more than repeated redesign. Nevertheless, at every level, decisions must be made which require some understanding of the systems as a whole, and the rationale of its implementation.

A central consideration is the objectives of the system or subsystem. Unless the objectives are clearly defined, the success or otherwise of an implementation remains unknown, and unless the objectives are relevant and clearly stated, the implementation is likely to be misdirected, and at worst, a complete waste of time. There are many possible reasons why a computer might be used in a particular application, such as: to save money, or staff time; to give more accurate or more repeatable results; to enable more information to be extracted from the data; to obtain results more

247

quickly; to make dissemination of the data or results more rapid; to achieve wider compatibility of data obtained from several sources; to gain prestige by pioneering a new method; or to gain expertise with a potentially important technique. All these and many other objectives can be appropriate in the right setting, but innovation for its own sake is unpopular in a project where the main aim is to obtain geological results in the minimum time, whereas rapid dissemination of results may be irrelevant to a research thesis.

The computer enthusiast is liable to be driven by hidden motives that are at odds with the objectives of the project. A geologist's personality may cause him to feel ill at ease with the complex diffuse ideas of geology, and lead him to seek sanctuary in the smaller, simpler world of computer programs that are within his complete control. Others may enjoy playing with new toys, or creating an atmosphere of impenetrable mystery that their colleagues cannot share. Certainly, it is a characteristic of many programmers that they are over-critical of programs written by others, and as objective in the assessment of their own work as a mother of her child or an author of his book. Fortunately, for most scientists, the greater satisfaction lies in developing an objective, professional skill, which enables them to respond, to the best of their abilities, to the task in hand. A good starting point in most applications is to list the possible objectives, perhaps rating them for relative importance, and to consider examples of how they could be met.

6.2. Review of the existing system and available resources

One effect of drawing up examples, may be to extend the possibilities beyond the area of immediate relevance. The next step may therefore be to look at the scope of the project. The extent to which a formal definition of the system is attempted depends on the circumstances. Generally, the larger the group, and the more complex the overall organization, the more precise must the definition be. If the geologist is responsible for only one part of a larger application, then the extent of the subsystem and its interfaces with the other subsystems are particularly important. If the project is an entirely new departure, then the computer involvement can be planned from scratch. Frequently, however, the requirement is to introduce computer methods in an existing activity. Information must then be collected about the present system. This might include an organization chart and list of staff responsibilities, a note of the types and amounts of data collected and recorded, an indication of the data flow, of how the data are processed at each stage and by whom, and the forms in which the information is eventually used, filed, and made more widely available. Estimates of the cost and staff time associated with each aspect of the existing system should be assembled and, if relevant, the time which elapses between various steps in a sequential process, such as collecting rock samples, submitting them for chemical analysis, receiving the results, mapping and interpreting the results. The way in which this is done must again be matched to the circumstances. The structured interviews and standardized documentation used by systems analysts in a commercial application might be totally inappropriate in an informal geological setting. However, even if the analysis is carried out by a geologist already familiar with existing procedures, it can be informative to obtain and tabulate some quantitative estimates for discussion.

By this stage in preliminary analysis, the scope and nature of the activity in which computer methods might be applied should be clear, the advantages to be sought from the computer should be known, and the analyst should be familiar with the existing system. A note might be made of ways in which the existing system could be improved without the computer, by revised manual procedures, or the introduction of new filing systems, edge-punched cards, desk calculators, facsimile transmission, microfilm or microfiche, etc. When the computer proposals are clarified, they can be compared, not only with the existing system, but with possible revisions of it. If it seems clear that there would be benefit in introducing computer methods, the next step could be to investigate more fully the available computing facilities. For most geologists working on a new project within a large organization, whether in industry, government or education, the computing facilities and policies will be already determined, and they have the major advantage of working within a defined framework with access to expert advice. A new project may require additional equipment and support, but experts will be available to advise and assist in obtaining suitable facilities, compatible with the organization's overall developments. In many smaller geological organizations, computer methods may be an entirely new departure, and require internal decisions on a wider range of topics. Nevertheless, it should be possible to obtain outside advice and support.

Consultants, service bureaux, computer manufacturers and others are generally willing to make their services available, at a price. Geological surveys, government agencies, and academic institutions can often obtain advice and assistance at less than the commercial rates, or even without charge, from other agencies or universities, from similar organizations abroad and, where appropriate, from foreign aid schemes. The value, though not necessarily the cost, of any external support is related to the level of specialist expertise, which can be very high in computer applications for the oil industry, including the processing of seismic data, log analysis, spatial analysis and display, automatic contouring, and banking of borehole data. Advice on computer installation and purchase, recruitment of computer staff, and software for statistical analysis, graphic display, image processing, data communications and text processing can also be readily obtained, but seldom with any direct understanding of the geological application. In general, it is desirable to retain control of the project, and ensure that the level of knowledge internally is sufficient to understand, assimilate and build on any information from outside sources. The exceptions tend to be in clearly defined specialized areas, like processing seismic records, where the procedures and results are fully understood, and where more than one source of external support can be found. The use of external facilities may also be appropriate for data preparation, computer time, digitizing and plotting, and use or support of software, since the control, application and interpretation remain with the geological project. Decisions in this area may be based on a straightforward comparison of estimated costs and convenience. Again, it might be desirable to ensure that a second supplier would be available if the first failed to come up to expectations.

A survey of available facilities can be helpful to the user in clarifying his own ideas, and should give him some knowledge of the availability and cost of relevant hardware, software and staff. Comparing notes with computer users in related fields is also likely to be helpful. Common sense indicates how far the survey should go. The cost of a simple application may be so low that even comparing prices is not worth while, whereas preparations for establishing an extensive data base may require many man-years of effort. One desirable result of the survey should be a clearer idea of what can be readily achieved on available facilities.

6.3. The system input and output

With preliminary ideas of the reasons for introducing the computer, of the project in which the applications will be made, and of the available facilities, more detailed attention can be given to the specific objectives, and the methods, benefits and costs of meeting them. The task is essentially one of matching available methods against the various operations. The results can be drawn up as a matrix or table where the rows list the operations, and the columns the computer methods, with symbols to indicate where the methods are appropriate and where they are likely to be most cost-effective. In a major project, several geologists will be involved, who may or may not be familiar with computer methods. As their cooperation is essential to the success of the operation, it may be advisable to assemble examples illustrating the computer input and output as seen by the geologist, and obtain their views on the value or otherwise of each application. If routine discussions can be established, the end product in more likely to meet the user's needs than a system designed in isolation by a computer expert.

In the examples, the input and output to each process assume major importance. This is entirely appropriate, as they are the interface between the user and the computer system. Unless they are acceptable to the geologist, it is unlikely that the system will succeed. Furthermore, although each operation may be regarded as a 'black box' by the user at this stage, the examples will help considerably to clarify the function of the black box. With most computer programs there is considerable flexibility in the form of data input and format of the results. It is not usually necessary to use real data during the initial planning, and an example using a test data set or taken from the literature may be more informative.

The input document which the geologist completes is at least one step removed from the computer, if the data entry is done by an operator. The design of the data sheet is a major topic, mentioned in §2.11. It may be desirable, in a new area of computer application, to design a sheet that provides two copies, one for a conventional manual file, with space for additional comments, and one for the data preparation section. Many attempts have been made to cut out data preparation as a separate activity, and record the data in the field in computer-readable form. In general, however, the second look which the geologist can give to his data in transcribing them for the computer record appears to be worth while, and in a new project, it is advisable to begin with more conventional forms of data input. The actual mechanics of data entry depend largely on the facilities available. Although punched cards are being gradually superseded for scientific data input, there may be little choice if the computer can only accept data in this form. Local terminals with cassette tape or diskette storage are cheaper than card punches, but rely on transmission to the central installation from the terminal or a compatible device. If more than one method is possible, a comparison of cost and convenience can easily be made. At this stage too, attention should be given to the codes that are to be used (see §2.14) and the thesauri required for vocabulary control (§2.15).

Experimentation with various designs of input and output document should precede the coding of the input and output procedures. For input, a hand-drawn data sheet can be reproduced on an office copier and should be entirely adequate for a trial run. Any necessary changes can be made before the final version is printed. The arrangement of computer output can be designed on a form specially made for this purpose with the same dimensions as output from a standard line printer. After a format has been defined, a typical set of output values can be inserted on a copy of the form, and a mock-up of the output prepared by entering the characters on punched cards or from a terminal and listing them on a line printer. The procedure can be repeated with any required improvements before the program is finally written. The output design form serves as a guide for writing the program, and if format numbers are listed on the form, it can become part of the permanent documentation of the output subroutine.

6.4. The driving forces

There are other aspects of the system to consider before the main work of programming begins. The user reaction to examples and mock-ups will give an indication of the acceptability of computer methods. But the objectives of the system are ultimately geological, and a complex system can work only if many different individuals play their part. Each component of the system operates because of a driving force. If the introduction of computer methods lowers the status of one individual, or creates unwanted extra work for another, it is illogical to expect their whole-hearted cooperation. Thought should therefore be given to the people involved, and how new methods will affect their way of working and their job satisfaction. Surveys of computer implementations in commercial activities have shown that a high proportion were seen in retrospect as failures. Those that succeeded were often characterized by a high degree of interest and involvement by the top management of the organization concerned. One recipe for success, therefore, is to make sure that the computer methods are introduced in a project where the project leader and higher management wish to see them succeed, and will take a close and informed interest in the techniques.

The systems analyst, for his part, should ensure that as far as possible he has a clear idea of the management objectives in the project as a whole, and the reasons and hopes for introducing computer methods. He can then consider the benefits that the various applications might bring in the light of these objectives. The question of whether the necessary incentives are there to drive the system, and the changes, including changes of job definition, that might be necessary to make the system work, should also be kept in mind.

6.5. Cost-effective implementation; hardware, software and methods

In the inevitable trade-off, the benefits of new methods are obtained at a cost. Having established what is possible, acceptable and useful, the next point to consider is what the methods will cost, and whether the cost will be justified by the benefits. A careful review of the proposals may indicate possible rationalizations, and may also point to comparatively minor operations within the computer activity that could be performed manually, with a major saving in the overall computer costs, perhaps because the operation could not be performed by standard programs or equipment.

In looking at the costs of an implementation, the three aspects of hardware, software and liveware—equipment, programs and staff—must be considered together since they are mutually dependent. A computer is of no value without adequate programs and staff to run it. Some aspects of the software were mentioned in §5.6, such as the desirability, to ensure adequate support, of concentrating on a few limited areas, of obtaining existing programs where possible, and of ensuring that standards are defined and maintained to allow an easy flow of data between the parts of the system. A major package in, say, data management or spatial analysis, may require one full-time member of staff to support and advise on the most effective use of the package. The cost of transferring a package to a new computer may amount to over 20% of the original development costs. If the full facilities are not needed, a simplified version or a simpler program offering a restricted range of options could be more cost-effective, and if it is always used in the same routine manner, would require very little support. Flexibility and the need to meet idiosyncratic requirements can increase software costs many times. The cost of writing software in-house is easy to underestimate. Although a skilled programmer can write a 50-line program in an afternoon, 80 to 90% of the overall effort lies in program design, testing, correction and documentation. A final production system of 10 000 lines of tested and documented code might require five to ten man-years of programming effort to complete.

With hardware as with software, there are advantages in turning to the services of existing computer installations. They will not be precisely tailor-made to the user's requirement, but the capital and running costs will be spread over many users. On the other hand, commercial computer bureaux have high overheads and are likely to be considerably more expensive than a well run and fully used in-house installation. An idea of the costs of using a bureau may be obtained by running 'benchmark' jobs which are considered to represent the kind of work that will be typical of the installation. The estimated number of production runs in an implementation may have to be increased by a factor of five to ten to allow for repeating failed runs, testing programs and selecting the most appropriate parameters.

An alternative to the computer bureau is the minicomputer, which offers considerable computing power at comparatively low cost. It might well be justified for a geological project with several users and a heavy computing load, particularly if a large computer is not conveniently available. It may be possible to split computing work between small tasks which are done interactively on an in-house minicomputer, and more extensive data management, large-scale statistical analysis and modelling which can be done in batch mode on a large external computer (see §5.5). If this approach is adopted, the interface between the two machines needs careful planning. It would probably involve exchange of magnetic tape as well as direct transmission, and a demonstration of software as well as hardware for these purposes should be requested before purchase, as it is not always as straightforward as it appears.

Minicomputers seem to succeed best at two levels. One is where they operate as an extended desk calculator, with additional memory, possibly including cassette tape or diskettes, and a faster printer, with some route for communicating data sets to and from a larger computer. The other level is a minicomputer with multi-user access, efficient compilers, some tens or hundreds of megabytes of random access memory, reasonably fast input and output devices, and good synchronous and asynchronous communications facilities. The amount of computing power that will ultimately be required on an implementation is easily under-estimated, and a computer which is too small for the job wastes both time and money. The computer must be adequate for the programs that are to be run on it. Converting software for a machine which is appreciably smaller or otherwise significantly different from the original may mean virtually rewriting the programs. A demonstration that a program will operate and give correct results for a test data set does not mean that the software can be used in a production environment. Indeed it is not impossible that the test data set contains the only data for which the program gives the correct results. With complex calculations, apparently valid results may be produced which are in fact incorrect due to a faulty algorithm, or to rounding errors related to word length. A cross check with a full data set, using another program and preferably a different method, perhaps on another installation, or a test using a published or simulated data set for which the correct results are known, may show whether the programs can be used on the chosen machine with some confidence.

If a need is clearly identified for purchasing hardware or software, or even the services of a bureau, the selection procedure tends to fall into a pattern. First, the literature, in particular the trade directories, can be searched to prepare a list of possible contenders, to which can be added suggestions from other organizations that have made similar purchases. Second, the vendors can be invited to send literature on their products. Third, a short list can be drawn up, paying particular attention to factors which rule out certain solutions, such as cost, incompatibility, the need for extensive modifications, inadequate support, unacceptable limitations in the hardware or software, inability to demonstrate existing implementations, uncertainty about the ability of the vendor to deliver the goods, or to provide future support or extensions to the system. These aspects will not be mentioned in the sales literature, and are likely to require some time on the telephone. At this stage information should be collected on not only the initial cost, but also the costs for evaluation, enhancement, maintenance, training, if any, and payment plans, if relevant. Fourth, presentations can be scheduled for the most promising systems, at which the product and the company behind it should be evaluated, and the representatives should be questioned on such aspects as delivery times, error reporting procedures for software, documentation and revisions, and the number and nature of existing installations. Final evaluation of the products still in the running can then be made, including reference, by visit or telephone, to existing installations, preferably not in the presence of the salesman. It is usually desirable for potential users to be represented in the later stages of the presentations and evaluation, and in the final purchase decision.

Not only must the programs be suitable for the available equipment and vice versa, but the staff must be capable of dealing with both. This may prove to be a limitation on the rate at which the computer applications can be developed, since staff with exactly the right qualifications are not likely to be available, and training takes time. The systems analyst function requires knowledge of the geological background, of available computer equipment and its advantages and disadvantages, and a detailed knowledge of the range of computer methods that are appropriate. As geological applications of the computer differ significantly from those in other fields, the systems analyst function will probably be best met by a geologist with a good background in computers. He will need to have, or to acquire, a working knowledge of the programming language selected for the application. The programming function can be handled by a programmer familiar with the language, even if he is not initially familiar with either geology or the application. He would then, however, require close supervision by the systems analyst. On a small project, the task of explaining the requirement to a programmer may be greater than that of writing the program, and the geologist or systems analyst could undertake the task themselves. On a larger project, however, a good programmer under careful supervision should produce more consistent, more efficient and better documented software than a group of geologists. Programming support can be obtained from service bureaux, and if only a limited number of packages will be used, which the bureau already supports, then this could be a good solution. Otherwise, it is normally desirable that the programmers should work in-house. With very large projects, programmer teams and structured programming are of interest, but these topics are outside the scope of this book. Training will be essential in any new application to ensure that skills reach the necessary levels in all fields, and courses of up to a week are available on many topics from computer manufacturers, software houses, bureaux and consultants. The need for training for both geologists and computer staff should be kept in mind in estimating staff costs and expected rates of progress. New staff are unlikely to be fully productive in their first year, and a significant proportion will not be productive at any time. The time of existing staff to train new recruits, and to train existing staff in new methods will have to be found, as well as personnel competent to undertake the training.

Data preparation is an area of major importance which may account for one-third to one-half of the total costs related to computing. If highly skilled data preparation operators with some years of experience can be found, they can possibly enter data from a terminal or on punched cards at ten to twenty thousand key depressions per hour. Whether they will continue to do so over a long period is another matter. Data will probably be entered a second time for verification, in order to reduce keying errors, and programs as well as data will have to be keyed. For various reasons, large segments of the data will probably have to be altered, re-entered and verified again. Having estimated the amount of data, it is not unreasonable to multiply by ten to estimate the amount of keying involved. If a less skilled operator is used, the keying rate may drop to a thousand key depressions per hour, with a high residual error even after verification.

Another hazard which can play havoc with the original estimates is the validation and correction of data. The difficulty seems to arise when a new data file is being established, in which, perhaps under pressure from the geologists to include all relevant data in case they prove to be useful, the input is highly complex, and probably highly encoded. The geologist, having laboriously encoded his data, believing that because they are potentially important they should be recorded on the computer, not unnaturally believes for the same reason that the computer record should be correct. Recording, encoding, transcription, keying and verifying all give opportunities for error. Add to this that as the project progresses, the geologist may change his own views on say, lithologies and stratigraphic markers that have already been entered. With several individuals concerned with the various stages of handling the data, the corrected material may have to pass through the same chain. Only a robust and flexible computer system and an alert and conscientious staff can hope to cope. It is not unknown for between four and ten correction cycles to be necessary for one set of data, nor for a correction phase to introduce more new errors than it corrects. In all probability, the final conclusion will be that there was no good reason for entering most of the data on the computer in the first place. A simpler system, working to defined and stable objectives, involving fewer people and less diffuse responsibility would have cost less, and probably produced useful results on time.

When a realistic appraisal is made of the costs and time taken to introduce computer methods in a short-term geological investigation, it may be decided that the benefits are not sufficient to justify full-scale computerization; that a geologist should undertake the systems analysis, because only he has sufficient motivation to make the system work; and that only a few programs should be used, with a corresponding limitation on the range of results, because the risk of delay in writing new programs is not acceptable.

The result might be a small success, which is certainly better than a big failure. Nevertheless, a realistic and hard-headed proposal should avoid undue timidity as well as over-ambitious initial plans. The costs may be high, but the benefits may be considerable. Oil and mining companies have to remain competitive with their rivals, government agencies have to be able to discuss, say, resource analysis on equal terms with the companies, and academic research and teaching should lead, not fall behind the state of the art. In the long run, failure to innovate may bear the higher cost.

The level and type of computer applications at which a project should aim is difficult to define other than in generalities. It should be driven by clear objectives and reasonable expectations of what the computer can produce as output, not by what is available as input. There are major advantages in having an overall plan of which individual implementations are a recognizable part, so that effort is not duplicated unnecessarily, and relative priorities can be decided. The plan should be flexible, recognizing the changing, dynamic nature of the system, and realizing that, as new applications are introduced, users will change their ways of working and their expectations of future support. The data, programs and equipment should as far as possible be adequate to deal with the full range of demands made on them. It is expensive to reorganize a data file because of new fields that have to be added which were not thought of originally, or to have to change to a completely new computer because demands had been greatly underestimated. Although planning may be comprehensive, the implementations need not be, and the introduction of new applications can wait until the requirement is established.

6.6. Organizational aspects

The need for the involvement of higher management has already been mentioned. It may be necessary to establish a separate computer section to support a set of major applications, if only to ensure that the overall computer needs of the organization are not overridden by the priorities of individuals. The computer section is likely to be concerned with the activities of the organization as a whole, although it may be staffed at a comparatively junior level. It is thus not easy to fit into the existing structure. A steering or advisory committee for the computer section, in which the relevant departments as well as the computer section are represented at a high level may help to ensure that the user groups are fully committed to the project.

On a major project it may be necessary to have an initial feasibility study to demonstrate that the various operations can be performed on the available facilities. This might be followed by a pilot project in which the full system is implemented, although not on a production scale, in order to discover any shortcomings in the methods, and to arrive at more accurate estimates of cost and time. The full-scale production system should follow on from this. If it involves major innovations, it is advisable to run the production system in parallel with a conventional system for a period, until the reliability of the production system has been fully demonstrated in practice.

After implementation, the working of the system should be reviewed at intervals, perhaps once a year, preferably by someone from outside the group responsible for the implementation. The usage and cost of computer time can be reviewed, as well as the growth of data, files and programs, the manpower time and costs of system design, programming, operations, data preparation and clerical work, error rates, and turn-round times at all stages. It should also be considered whether the objectives are being met, whether the computer output is used as expected, whether the superseded manual methods have in fact been discontinued, whether standards are adhered to and whether documentation is satisfactory. Overall, the review should consider whether the original objectives are being met at a reasonable cost, and whether they require to be changed or revised in the light of experience. The views of the users as well as the computer staff are obviously important, and their knowledge of related activities outside the immediate area of application may provide opportunities to expand or rationalize the original concepts.

The most significant recent advances in computer applications in geology may well be the major systems of software and data resulting from the combined efforts of organized groups. Many future developments, on the other hand, may stem from the initiative of individuals who can base their own research on the foundation of these large-scale projects. The availability of cheap minicomputers and of good computer communications can give the geologist control of his own activities while giving full access to data and systems maintained by others. Developing areas of research, such as computer modelling, give scope for a wide diversity of approaches on every scale. As was the case when computer methods were first introduced, therefore, it may be individuals working on their own, unhampered by organizational constraints and free to follow their own ideas, who first map out the paths that will be followed by the next generation of computer methods in geology.

6.7. Recapitulation

It is desirable in any computer application to keep in mind the activities which the geologist can perform much better than any computer. The background geological knowledge, the ability to detect and recognize complex patterns and to visualize past processes and sequences of events from the scanty evidence of present-day boreholes and outcrops, are essential aspects of geological investigations. They rely entirely on the human mind. Computer methods can extend the reach of the mind and the ability to share information. Retrieval of relevant references from a bibliographical data base is a simple example. More fundamentally, the systems approach offers a philsosophy and a set of techniques for formulating and implementing a computer application. It should lead to a more rigorous description of a system, with the subsystems, components and interfaces defined and described in order to clarify the overall objectives and mechanisms, and to indicate the possibilities and consequences of introducing computer methods, within a modular framework which makes it possible to match the method to the requirement. The power of the systems approach stems from the concept of quantitative measurement; the flexibility of transmitting, processing and presenting information held in digital form; the power of the computer in calculation and data management; and the ability to express concepts in a general form through mathematics and computer languages.

Quantitative measurement provides a portable yardstick for comparing and ordering observations to any degree of precision. An exact operational definition, which spells out the procedure used in measurement or classification, ensures that comparable results can be obtained by all individuals using the same procedure. Digital representation of information in computer and communications systems, in which measurements, text and graphic information are represented by patterns of discrete pulses of electricity or patterns of magnetic flux, makes it possible to transmit, store, select, manipulate, transform and analyze information without any undesirable loss of precision. Text and numeric data can be transformed into digital form from a keyboard terminal and graphic data from a digitizing table. Data transmission is possible to most points that can be reached by telephone and the digital data can be presented in printed or graphic form on a wide range of devices from a line printer to a typesetter or a microfilm camera, and from a visual display unit to a cartographic plotter.

Programming languages make it possible to define a series of operations including input, transmission and output of data in a form which the computer can accept, and a programmer readily understands. Similarly, mathematical and data management operations can be expressed as a sequence of computer instructions. Programming languages are generally hierarchical, in the sense that one statement generates many machines language instructions, one subroutine contains many statements, one program may call many subroutines, and one system may contain many programs. They are modular, in the sense that the same subroutines can be used to build many different programs. The computer program has two roles, as a sequence of instructions to the computer, and as a means of describing a procedure, a method of analysis or a model which other geologists can understand and use.

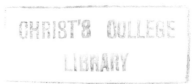

Where data are recorded for computer input, it may be necessary to work to carefully defined conventions. Recording data on a well designed data sheet with appropriate codes and word lists may help to reduce errors and ensure that the data records are comparable, unambiguous and uniform. The sampling methods must be appropriate to the uses to be made of the data. Data management, which involves the input, checking, correction, sorting, storage, editing, updating, deleting, amending, selection, retrieval and output of data may be regarded as a separate activity to data analysis. It is a major requirement in the creation and operation of data banks, which may have in various degrees the attributes of a data library, a data base and a computable model.

Mathematics provides a language in which concepts can be expressed without restriction to any specific application, thus giving solutions of general application. Data analysis methods of proven value in geology include calculation of descriptive statistics, cluster analysis, correlation, principal component analysis, contingency tables, analysis of variance and regression. Some are particularly relevant in their application to spatial data in such techniques as autocorrelation, transition probabilities and trend surface analysis. The concepts of probability theory lead to methods for quantifying uncertainty and risk, and for building simulation models from which the outcome of complex sequences of events can be deduced. Graphic display of the raw data or the results of data analysis or simulation presents the information in a form which the geologist can more readily assimilate. The systems approach can lead to the development of computer models with different topics, scope and detail, and a consideration of how to resolve discrepancies between them. Digital networks provide a means of exchanging and transmitting information in text and graphic form, and may lead to more effective exchange of ideas and use of expert opinion by such techniques as teleconferencing and Delphi methods.

The convergence of technologies of computing, text and graphics handling, and digital transmission may encourage complex, integrated developments on a scale which only large organizations can assimilate. They are, however, frequently available on a service basis to the individual user, and provide well established techniques on which the scientist can build to meet his own specific requirements. The text books now available in this field, and recent issues of the periodicals specializing in geological applications of the computer (see references) give guidance and examples of the chosen methods. A suitable overview of the field, to determine which methods are appropriate and how full advantage can be taken of the available services, has already been obtained by the reader who has arrived sequentially at this point.

REFERENCES

The list has been kept short. The cited text books are a good source of further information and provide many additional references. The works by Davis (1973) and Harbaugh *et al.* (1977) are a good entry point into the mainstream of geological applications. More recent material can be found in the journal "Computers & Geosciences", or by searching a bibliographical data base with an appropriate profile. Other references are included to illustrate or substantiate particular points in the text, to call attention to papers of particular relevance, or to identify the source of various concepts. They are not intended as a "reading list".

Ahuja, D. V. and Coons, S. A. (1968). *IBM Systems Journal* 7, 188–205.

Brisbin, W. C. and Ediger, N. M. (eds.) (1967). "A National System for Storage and Retrieval of Geological Data in Canada". Geological Survey of Canada, Ottawa.

Chandor, A. (1970). "A Dictionary of Computers". Penguin Books, Harmondsworth.

Codasyl (1971). "Data Base Task Group Report". Association for Computing Machinery, New York.

Coan, D. R. A. and Sharratt, J. R. (1970). "Programming Standards, 1: Documentation". National Computing Centre Ltd., Manchester.

Codd, E. F. (1970). *ACM Commun.* 13, 377–387.

Craig, G. Y. (1969). *Scott. J. Geol.* 5, 305–321.

Craig, G. Y. and Oertel, G. (1966). *Quart. J. Geol. Soc. London* 122, 315–355.

Cutbill, J. L., Hallan, A. J. and Lewis, G. D. (1971). *In* "Data Processing in Biology and Geology" (J. L. Cutbill, ed.), pp. 255–274. Academic Press, London and New York.

Daniels, A. and Yeates, D. (eds.) (1969). "Basic Training in Systems Analysis". Sir Isaac Pitman, London.

Davis, J. C. (1973). "Statistics and Data Analysis in Geology". Wiley, New York.

Deen, S. M. (1977). "Fundamentals of Data Base Systems". Macmillan, London.

Dixon, W. J. and Massey, F. J. (1957). "Introduction to Statistical Analysis". McGraw-Hill, New York.

Farmer, D. G. and Read, W. A. (1976). *Computers & Geosciences* 2, 365–374.

Fisher, R. A. (1953). *Proc. Roy. Soc., London, Series A* 217, 295–306.

Forrester, J. W. (1961). "Industrial Dynamics". Wiley, New York.

Forbes, C. L., Harland, W. B. and Cutbill, J. L. (1971). *In* "Data Processing in Biology and Geology" (J. L. Cutbill, ed.), pp. 311–320. Academic Press, London and New York.

Good, D. I. (1964). *Kansas Geol. Survey Bull.* 170, part 3.

Gover, T. N., Read, W. A. and Rowson, A. G. (1971). *Inst. Geol. Sci. Rept.* No. 71/13.

Harbaugh, J. W., Doveton, J. H. and Davis, J. C. (1977). "Probability Methods in Oil Exploration". Wiley, New York.

Harbaugh, J. W. and Bonham-Carter, G. (1970). "Computer Simulation in Geology". Wiley, New York.

Harbaugh, J. W. and Merriam, D. F. (1968). "Computer Applications in Stratigraphic Analysis". Wiley, New York.

Jeffery, K. G. and Gill, E. M. (1976). Computers & Geosciences 2, 345–346.

Jones, T. A., Baker, R. A. and Dumay, W. H. (1976). Computers & Geosciences 2, 351–355.

Kendall, M. G. (1961). "A Course in Multivariate Analysis". Charles Griffin, London.

Koch, G. S. and Link, R. F. (1970). "Statistical Analysis of Geological Data". Wiley, New York.

Krumbein, W. C. and Graybill, F. A. (1965). "An Introduction to Statistical Models in Geology". McGraw–Hill, New York.

Lea, G., Charles, R. and Shearer J. (1973). "Geosaurus: Geosystems Thesaurus for Geoscience". Geosystems Publications, London.

Leontief, W. (1966). "Input-Output Economics". Oxford University Press.

Linstone, H. A. and Turoff, M. (eds.) (1975). "The Delphi Method: Techniques and Applications". Addison–Wesley, Reading, Mass.

Loudon, T. V. (1964). Tech. Rept. 13 of ONR Task No. 389–135. Northwestern University, Illinois.

Loudon, T. V. (1971). In "Data Processing in Biology and Geology" (J. L. Cutbill, ed.), pp. 135–145. Academic Press, London and New York.

Loudon, T. V. (1974). Inst. Geol. Sci. Rept. No. 74/1.

McCracken, D. D. (1965). "A Guide to Fortran IV Programming". Wiley, New York.

Miller, R. L. and Kahn, J. S. (1962). "Statistical Analysis in the Geological Sciences". Wiley, New York.

National Computing Centre Ltd. (1970). "Standard Fortran Programming Manual". National Computing Centre, Manchester.

National Computer Centre Ltd. (1970). "Systems Documentation Manual". National Computing Centre, Manchester.

Nie, N., Bent, D. H. and Hull, C. H. (1970). "Statistical Package for the Social Sciences". McGraw–Hill, New York.

Oppenheimer, C. H., Oppenheimer, D. and Brogden, W. B. (eds.) (1976). "Environmental Data Management". Plenum Press, New York.

Pettofrezzo, A. J. (1969). "Matrices and Transformations". Prentice–Hall, Englewood Cliffs.

Piper, D. J. W., Harland, W. B. and Cutbill, J. L. (1971). In "Data Processing in Biology and Geology" (J. L. Cutbill, ed), pp. 17–38. Academic Press, London and New York.

Plant, J., Jeffery, K. G., Gill, E. M., and Fage, C. (1975). J. Geochem. Expl. 4, 467–486.

Raup, D. M. (1966). J. Paleontology 40, 1178–1190.

Schwarzacher, W. (1975). "Sedimentation Models and Quantitative Stratigraphy". Elsevier, Amsterdam.

Stoll, R. R. (1961). "Sets, Logic and Axiomatic Theories". W. H. Freeman, San Francisco.

Stuart, A. (1962). "Basic Ideas of Scientific Sampling". Charles Griffin, London.

Till, R. (1974). "Statistical Methods for the Earth Scientist. An Introduction". Macmillan, London.

Vallee, J. and Wilson, T. (1976). *Computers & Geosciences* **2**, 305–308.
Watson, G. S. (1966). *J. Geology* **74**, 786–797.
Yule, G. U. and Kendall, M. G. (1958). "An Introduction to the Theory of Statistics". Charles Griffin, London.

SUBJECT INDEX

Abbreviations, 92ff
Absent data, 77, 88, 97
Accession numbers, 93
Acoustic couplers, 234
Address, **32**, 37, **40**, 90
Affine transformations, 145
ALGOL, 38
Allocation models, 200ff
Analysis of variance, 4, 73, 156, 157ff, 191
Ancillary equipment, 27
And (in formal logic), 31, **57**
Anomalies, 73, 126, 152–154
ANSI, 38, 48
APL, 233, 238
Argument lists, 44, 45
Arrays, **40**, **58**ff, 79, 83
ASSEMBLER, 38, 45
Asynchronous transmission, 235
Autocorrelation, 170, 190, 206
Autoregression, 172
Auxilliary equipment, 27
Averages (*see* Mean)

Backup, 99, 104
Balanced systems, 21, 78, 176
Bar charts (*see* Histograms)
BASIC (language), 38, 233
Batch processing, 231ff
Benchmarks, 253
Biased statistics, 129
Bibliographical data, 241, 244
Bimodal distributions, 126, 131
Binary arithmetic, 30ff
Binary-coded decimal (BCD), 30ff, 33
Binary units (Bits), 30ff, 96
Bistable state devices, 28, 29, 31
Bits, 30ff, 96
Black boxes, 217, 251
Branches (in program), 42

Buffers, 35
Bytes, 31ff, **32**, 88

Cafeteria systems, 232
Call statements, **44**, 106
Canonical regression, 214
Card punches (*see also* Punch cards), 16ff, 33
Cartography, 138, 146
Cassette tapes, 21ff, 36, 234
Catalogs, 244
Cause and effect, 149, 176, 189
CCITT, 233
Central Limit Theorem, 187
Central Processing Unit (CPU), 14
Central tendency, 180
Chi-squared test (χ^2), 189
Circuits, 31
Classification models (*see also* Cluster analysis), 212
Closed-number systems, 62
Cluster analysis, 4, 118, 120, 130ff
COBOL, 106
Codasyl, 105
Codes, 83, 92ff, 104, 198ff, 251
Comments file, 85
Common statements, **44**, 45
Communication (see also Data transmission), 11, 36, 54, 65, 227ff
Compartment models, 217ff
Compatibility, 21ff, 22, 38, 52
Compilers, 38, 39, 40
Conditional branches, 42, 43
Conditional probability, 182
Confidence limits, 71, 186
Constants (in FORTRAN), 40
Contingency tables, 155, 156, 168
Contouring, 12, 164, **202**ff, 223
Coordinates, 25, 114, 139, 145
Core, **29**, 30
Correlation, 4, 133ff, 155, 188, 189

265

Correlograms, 170, 171
Cost-benefits, 252
CPU (*see* Central Processing Unit)
Cross-references, 83, 90, 92, 102, 198ff
Current awareness, 245
Curve fitting, 151
Cybernetics, 214ff

Data administrators, 99
Data analysis, 4, 13, 109ff
Data banks, 65, 99ff
Data bases, 95, 99ff
Data collection, 10, 53ff, 138
Data descriptions, 100, 101, 105, 106
Data library, 100ff
Data management (*see also* Data retrieval), 13ff, 91, 99, 102, 106
Data manipulation language, 106
Data preparation, 14, 15ff, 53ff, 78, 79ff, 87, 251, 256
Data retrieval, 56, 83, 90, 97, 218
Data sheets (*see* Form design)
Data structures, 58ff, 78, 99, 103
Data transmission, 22, 30, 36, 233
Data types, 62ff
DBMS (*see* Data management)
Default options, 51
Delphi methods, 222, 242ff
Dendrograms, 116, 118, 132
Desk calculators, 36, 254
Deviations (*see* Anomalies)
Dictionary files (*see also* Table lookup), 101, 105
Digitizing (of graphic data), 25ff, 34, 146
Direct data entry, 21ff
Direction cosines, 147, 148, 173
Direction ratios, 148
Discontinuities, 205, 206
Discriminant analysis, 133, 138
Diskettes (*see* Floppy disks)
Disks, 27, 28, **29**, 38
Display (*see* Graphics)
Distance, 113, 132
Distortion corrections, 145, 146
Distributed processing, 236
Documentation, 238ff
Dot raster printers, 33, 34
Double precision, 32
Drum storage, 29
DYNAMO, 215

EBCDIC, 16
Efficient sampling, 66
Efficient statistics, 129
EIA, 233
Eigenvalues, 137
Eigenvectors, 137
Encryption, 237
Epidemic models, 219
Error checking, 15ff, 19, 73, 94, 126, 149, 237
Euclidean distance, 115
Exchange formats, 88
Executive programs, 35, 52, 241
Experimentation, 8, 157, 177, 191, 196
Explanation, 10, 109, 175ff
Expressions (in FORTRAN), 40ff

F test, 190
Facsmile transmission, 83, 249
Factor analysis, 4, 138
Feasibility study, 257
Feedback, 215, 218
Fields, **19**, 58, **88**
Files, **86**, 88, 99
Filtering, 166, 218
Finite-element models, 221
Fixed formats, 47, 89
Floating point, 32
Floppy disks, 22, 36, 234
Flowcharts, 49ff, 60, 183
Forests, 60
Form design, 86–88, 251
Formats, 45, 47ff, 79, 88ff, 93, 241, 251
FORTRAN, 38, 39ff, 79, 97, 106
Frames (for data transmission), 237
Frames of reference, 3, 110ff, 121, 123, 147, 152, 211
Franklin Code, 95
Free formats, 48, 89, 95
Frequency distributions, 71, 121ff, 129, 173, 178, 179, 184ff

Generated codes (*see* Hash codes)
Geographic transformations, 164
Geometric data processing, 138ff
GPSS, 216
Graph plotters (*see* Plotters)
Graphics, 2, 6, 25ff, 123, 138ff, 160ff, 167ff, 233
Grey-scale, 165, 166
Ground truth, 213
Growth models, 220

Handwriting readers, 24
Hardware, 12ff, 14, 37
Hash codes, 95, 105
HASP, 234
HDLC, 234
Heuristic methods, 201, 210, 212ff
Histograms, 121, 122, 123ff, 185
Homogeneous coordinates, 145
Homogeneous populations, 126, 132, 133, 154
Hypothesis testing, 189ff, 195

IF statements, 42
Incompatibility (see Compatibility)
Indexes, 83, 85, 244
Information (see Data)
Information flow, 229
Inhospitable arrays, 93, 96
Input, 14, 45ff, 88, 250
Input–output analysis, 219
Instructions (in FORTRAN), 37, 39
Integers, 32
Interactive computing, 213, 231ff
Interfaces, 37, 38, 99, 217, 219, 227ff, 230, 231, 250
Interpretation, 11, 56, 109ff
Intersection, 55
Inverted files, 90, 105
ISO, 19, 20, 38

Job Control Language, 51

Key to disk, 21ff
Keyfields, 95, 98, 107
Keypunch (see Card punch)
Keywords, 98
Kriging, 171
Kurtosis, 129

Labels, 89, 101
Linear programming, 200ff
Line printers, 33
Location (in store), 32, 39, 40
Logarithmic transformation, 127
Loggers (data loggers), 24
Logic, 55, 56ff
Logic elements, 31, 37
Loops, 42, 59

Machine languages, 37
Machine translation, 5
Magnetic tapes, 15ff, 19, 27ff, 28, 33
Mainframes, 27
Mapping, 60, 112
Mark-sense cards, 24
Marketing (of software), 240
Markov chains, 209
Match coefficients, 118
Matrices, 58ff, 120, 132, 133, 136, 139ff, 220
Means, 71, 121, 184
Medians, 184
Memory, 27ff, 29, 39
Menu, 52, 233
Message concentrators, 234
Metrics, 114, 120, 128, 147
Microfilms, 35, 83, 249
Minicomputers, 27, 36, 51, 234, 253, 254, 258
Missing values (see Absent data)
Mnemonic codes, 38, 52, 85, 94, 198
Mock-ups, 252
Modes, 184
Models, 6ff, 8, 9, 71, 76, 77, 108, 149, 177, 181, 191, 197ff
Modems, 233, 234
Modularity, 52
Monte Carlo methods, 225
Moving averages, 164, 165
Multiplexors, 234
Multiprogramming, 35

Networks, 36, 60, 61, 90, 233ff
Noise, 112, 137, 166
Non-linear mapping, 138
Normal distributions, 71, 121, 122, 127 184, 185, 194
Normalization (see Standardization)
Normative analysis, 200
Notations, 54, 57, 58, 107, 209
Null hypothesis, 187ff, 196
Numerical analysis (see Data analysis)

Objective functions, 200
Observation, 10, 53ff
Offline, 27
Online, 27
Operational definitions, 259
Operating systems, 35, 36, 50ff

Optimization, 22
Or (in logic), 31, **57**
Orientations, 63, 137, 147ff, 171, 173
Output, **33**, 45ff, 250

Packages, 51
Paper tapes, 15ff, **19**
Parallel running, 258
Parity, 19
PCA, 4, 135, 149, 150
Peripherals, 27
Pilot projects, 257
Pixels, 212
PL–1, 38, 106
Plotters, **34**, 165
Plotting (*see* Graphics)
Pointers, 90
Populations, 64
Prediction, 180, 194ff
Principal component analysis (*see* PCA)
Printers, 33, 145
Probabilities, 121, 178ff, 185, 223, 224
Problem-oriented languages, 37ff
Process control, 5
Processors, 14
Process–response models, 214ff
Profiles, 245
Program libraries, 51
Programming strategy, 240
Projection, 145
Properties, 56, 58
Protocols, 227ff
Pseudo-random numbers (*see* Random numbers)
PTT's, 233
Punch cards, 15ff, **16**, 35, 41, 251
Purchasing procedures, 254

Qualitative methods, 62, 118, 154ff, 167, 168
Queues, 15, 35ff, 219

r (*see also* Correlation), 133
Random numbers, 191ff, 196
Randomness, 111, 121, 137, 150, 178ff, 184ff, 191ff
Ranked data, 62
Rasters (*see also* Dot raster printers), 27
Redundancy, 15ff, 19, 87, 103, 105, 111ff, 135, 137

Referral centers, 101
Registers (in a processor), 32
Regression, 4, **149**ff, 191, 214
Relational data bases, 107
Relationships, 56ff, 58, 61ff, 87, 103, 105, 107, 112, 133
Remote job entry, 234
Remote sensors, 212
Reservoir modelling, 221
Resource estimation, 221ff, 228
Response times, 232
Retrospective searches, 245
Rings (data structure), 61
Risk analysis, 178ff, 186, 221ff, 225
RJE, 234
Rotations (in geometry), 136, 138, **139**ff

Sampling, 56, 64ff, 178
  efficient, 67
  grid, 68, 74, 170
  multilevel, 70
  random, 68, 69, 71, 186
  sequential, 71, 72
  stratified, 70
Scaling (*see* Weighting, Stretching)
Scenarios, 191ff, 196
Schemas, 105, 238
Scientific method, 9ff
SDI, 245
Search methods, 65
Selective dissemination of information, 245
Semantic codes, 96ff, 198
Semiconductors, 29, 31
Sensitivity analysis, 196
Sequential data, 28, 159ff, 167ff, 170ff, 208ff, 211
Serial correlation, 170
Series analysis (*see* Sequential data)
Sets, 55ff, 105
Shared coding schemes, 227, 237, 238
Shearing, 144
Similarity functions (*see also* Distance), 119
Signals, 112, 166
Silicate recalculation, 200
SIMSCRIPT, 216
SIMULA, 216
Simulation, 65, 66, 178ff, 191ff, 195, 196, 214ff

Skewness, 129
Slope analysis, 173ff, 206
Smoothing, 166
Software, **14**, 37ff
Software houses, 106
Sparse arrays, 60
Spatial analysis, 159ff, 164
Spatial models, 202ff, 211ff, 223
Spectral analysis (*see also* Sequential data), 172, 206
Splines, 207
Stability, 215
Standard deviation, 72, 122, 185
Standard error, 186
Standardization (of data), 118, 122ff, 132, 133, 135, 150
Standards, 38, 94, 98, 105, 227ff, 2̄3̄0̄, 233, 235
Star network, 235
Statements (in FORTRAN), 41
Statistics (*see also* Data analysis), 4, 64, 110, 121, 129, 178
Steering Committees, 257
Stereograms, 173
Stop words, 98
Storage, 33
Stratigraphic correlation, 210
Stretching (change of scale), 142ff
Strings, 58, 85, 97
Subschemas, 105, 238
Subroutines, 39, **43**ff
Subscripts (of array) 40, 58
Subsets, 55ff
Subsystems, **8**, 10, 53
Symbols, 12, 13
Synchronous transmission, 234
Systems, 6ff, **7**, 9, 14, 197ff, 247ff
Systems analysis, i, 6, **7**, 81, 86, 198, 228, 242, 247ff

Table lookup (*see also* Cross-references), 85, 96, 198
Tapes (*see* Magnetic tapes, Paper tapes)
Technological gatekeepers, 230
Teleconferences, 222, 242ff
Terminals, 22, 35, 36, 233ff, 251
Text information, 5, 79, 92, 98, 242
Thesaurus, **98**, 104, 251
Topology, 55
Trade-offs, 21ff, 52, 88, 89, 229, 252
Transformations, 4, 112ff, 127, 128, 136, 182, 194, 198, 199
Transitions, 167ff, 208ff
Translation (shift of origin), 139
Transmission (*see* Data transmission, Networks)
Trees (data structure), 59, **60**, 90, 91
Trend surfaces, 152, 166, 188, 195, 206
Trigonometric series, 151, 171
Truncated distributions, 97
Typesetters, 35

Union of sets, **55**, 57
User manuals, 239
User-oriented languages, 37
Utility programs, 51

Variables (in FORTRAN), **40**, 58, 121
Variance, 71, 72, 122, 136, 153
Vectors, 58, 63, 148
Verification (*see also* Error checking), 19
Visual display units (VDU's), 22, 35, 233, 235
Vocabulary control, 98ff

Weighting, 74, 77, 118, 119ff
Word processing (*see also* Text information), 242ff
Words, 31ff, 32

x–y plots, 134, 135, 150, 153

R